Edward Emanuel Klein

The Bacteria in Asiatic cholera

Edward Emanuel Klein

The Bacteria in Asiatic cholera

ISBN/EAN: 9783744750349

Printed in Europe, USA, Canada, Australia, Japan

Cover: Foto ©berggeist007 / pixelio.de

More available books at **www.hansebooks.com**

THE BACTERIA

IN

ASIATIC CHOLERA

BY

E. KLEIN, M.D., F.R.S.,

LECTURER ON GENERAL ANATOMY AND PHYSIOLOGY IN THE MEDICAL SCHOOL OF
ST. BARTHOLOMEW'S HOSPITAL; PROFESSOR OF BACTERIOLOGY AT THE
COLLEGE OF STATE MEDICINE, LONDON

London

MACMILLAN AND CO.

AND NEW YORK

1889

The Right of Translation and Reproduction is Reserved

TO

GEORGE BUCHANAN, ESQ., M.D., F.R.S.,

MEDICAL OFFICER OF THE LOCAL GOVERNMENT BOARD,

THIS VOLUME IS

Respectfully Dedicated

BY

THE AUTHOR.

PREFACE.

THE present volume is a reprint of a series of articles published in the *Practitioner* (October 1886, to May 1887). Since that time a considerable number of contributions have been made to the knowledge of the comma-bacilli of Koch. These have been added to the present volume; but besides these no noteworthy addition has been made to our knowledge of the bacteria in Asiatic cholera. This need not at all surprise us, considering that with few exceptions most Continental pathologists consider the comma-bacilli of Koch as being the cause of cholera, that is to say as being the real cholera microbes. They consider the chapter of the etiology of *Cholera Asiatica* as closed, and there is therefore no need to look for any other and new cholera microbe. I have ventured to differ from this opinion when writing the above articles in the *Practitioner*, and after all the observations published by various pathologists since then, I still differ from the proposition that Koch's comma-bacilli have been satisfactorily proved to be the cause of cholera.

As then, so also now I hold that the comma-bacilli of Koch do not fulfil the conditions which the cholera microbe ought to fulfil. As then, so also now I agree with the prevailing opinion that the comma-bacilli of Koch are an important diagnostic guide.

I have been credited by various writers with the contrary statement; I have in the articles, as they first appeared, tried to correct such statements, though without effect, and I must therefore again state, that I fully agree with Koch and others as to the constant presence of the comma-bacilli in the cholera intestine and cholera discharges during the early stages, but that here our agreement ends. That the comma-bacilli of Koch are not accepted as the proved cause of cholera, in this I am not alone, as, for instance, Professor Baumgarten seems to think; von Pettenkofer, acknowledged to be the greatest living authority on the etiology of cholera, holds this view, viz. that the comma-bacilli are not the proved cause of cholera; Dr. D. D. Cunningham of Calcutta, who during many years had exceptional opportunities of studying this disease, is of the same opinion. The English Cholera Commission (Professor Roy, Dr. Sherrington and Dr. Brown); Dr. Shakespeare of Philadelphia, and Dr. v. Emmerich of Munich, are the most noteworthy observers who have arrived at the same conclusion after special study of cholera.

E. KLEIN.

CONTENTS.

	PAGE
INTRODUCTION	1

CHAPTER I.

THE HISTORY OF THE COMMA-BACILLUS	5

CHAPTER II.

THE DISTRIBUTION OF COMMA-BACILLI	22

CHAPTER III.

MORPHOLOGY OF THE CHOLERAIC COMMA-BACILLI	40

CHAPTER IV.

CHARACTERS OF THE COMMA-BACILLI IN ARTIFICIAL CULTIVATIONS	61

CHAPTER V.

VARIOUS SPECIES OF COMMA-BACILLI 82

CHAPTER VI.

DIAGNOSTIC VALUE OF CHOLERAIC COMMA-BACILLI 112

CHAPTER VII.

EXPERIMENTAL PRODUCTION OF CHOLERA 119

CHAPTER VIII.

THE INFECTIVENESS OF CHOLERA 141

CHAPTER IX.

OTHER BACTERIA IN CHOLERA 166

LIST OF ILLUSTRATIONS.

FIG.   PAGE

1. From a Preparation of Fresh Mucus-Flakes from a Choleraic Evacuation 24

2. Preparation of Mucus-Flakes from the Ileum of an Acute Case of Cholera 25

3. Preparation of Mucus-Flakes from the Ileum of an Acute Case of Cholera 26

4. From a Preparation of Fresh Mucus-Flakes from the Lower Part of the Ileum of a Typical rapidly Fatal Case of Cholera 27

5. From a Section through the Ileum of an Acute Case of Cholera 31

6. From a Preparation of Mucus-Flakes from the Lower Ileum, which had been allowed to undergo Putrefaction for three days. 42

7. Preparation of Choleraic Comma-Bacilli stained with Gentian-violet, and afterwards well washed 45

LIST OF ILLUSTRATIONS.

FIG.   PAGE

8. ARTIFICIAL CULTIVATION OF CHOLERAIC COMMA-BACILLI IN ALKALINE PEPTONE GELATINE. 47

9. ARTIFICIAL CULTIVATION OF THE SAME COMMA-BACILLI AS IN PRECEDING FIGURE IN ALKALINE PEPTONE BROTH GELATINE 48

10. ARTIFICIAL CULTIVATION OF THE SAME COMMA-BACILLI IN ALKALINE BEEF BROTH 49

11. FROM A CULTIVATION OF CHOLERAIC COMMA-BACILLI IN LIQUEFIED GELATINE, AFTER SEVERAL WEEKS . 51

12. PREPARATION OF A PURE CULTIVATION OF CHOLERAIC COMMA-BACILLI IN AGAR-AGAR MEAT-EXTRACT PEPTONE, SEVERAL MONTHS OLD 53

13. PREPARATION OF A CULTIVATION OF CHOLERAIC COMMA-BACILLI IN EGG-ALBUMEN AND AGAR-AGAR, TEN DAYS OLD 54

14. FROM AN ARTIFICIAL CULTIVATION OF CHOLERAIC MUCUS FLAKES ON DAMP LINEN 55

15. PREPARATION OF A CULTIVATION OF CHOLERAIÇ COMMA-BACILLI ON DAMP LINEN AFTER THIRTY-SIX HOURS 56

16. FROM A RECENT ARTIFICIAL CULTIVATION OF CHOLERAIC COMMA-BACILLI IN ALKALINE AGAR-AGAR JELLY. . 57

17. FROM AN ARTIFICIAL CULTIVATION OF CHOLERAIC COMMA-BACILLI ON NEUTRAL AGAR-AGAR JELLY AT ORDINARY TEMPERATURE, AFTER A FEW WEEKS. . 58

LIST OF ILLUSTRATIONS.

FIG.		PAGE
18.	FROM A SIMILAR PREPARATION	58
19.	FROM A SIMILAR PREPARATION	58
20.	PLATE-CULTIVATION IN GELATINE FORTY-EIGHT HOURS OLD, SHOWING YOUNG COLONIES OF CHOLERAIC COMMA-BACILLI.	62
21.	PLATE CULTIVATION WITH COLONIES OF CHOLERAIC COMMA-BACILLI SEVENTY-TWO HOURS OLD	66
22.	FROM A CULTIVATION OF CHOLERAIC COMMA-BACILLI IN GELATINE IN A GLASS DISH FOUR DAYS AFTER INOCULATION IN SPOTS	67
23.	PLATE CULTIVATION OF CHOLERAIC COMMA-BACILLI IN GELATINE AFTER FOUR DAYS AT 19°C	68
24.	A COLONY OF CHOLERAIC COMMA-BACILLI IN GELATINE, SEVENTY-TWO HOURS OLD.	71
25.	CULTIVATIONS OF CHOLERAIC COMMA-BACILLI	74—77
26.	STABCULTURES OF CHOLERAIC COMMA-BACILLI IN NUTRITIVE GELATINE AFTER ONE, THREE, FOUR, FIVE, SEVEN, AND TEN DAYS RESPECTIVELY.	79
27.	COVER-GLASS SPECIMEN OF FINKLER'S COMMA-BACILLI FROM A GELATINE CULTURE	83
28.	GELATINE PLATE CULTIVATION OF FINKLER'S COMMA-BACILLI AFTER INCUBATION FOR FORTY-EIGHT HOURS AT 20°C	86

LIST OF ILLUSTRATIONS.

FIG.   PAGE

29. CULTIVATION OF FINKLER'S COMMA-BACILLI IN NUTRITIVE GELATINE (10 PER CENT.) AFTER FOUR DAYS' INCUBATION 88

30. COVER-GLASS SPECIMEN OF MUCUS-FLAKES FROM A MONKEY SUFFERING FROM DIARRHŒA 91

31. COVER-GLASS SPECIMEN OF CONTENTS OF CÆCUM FROM A NORMAL GUINEA-PIG 93

32. COVER-GLASS SPECIMEN FROM A CULTIVATION IN 10 PER CENT. NUTRITIVE GELATINE OF THE NON-LIQUEFYING VARIETY OF COMMA-BACILLI FROM A CASE OF NOMA IN A CHILD 97

33. PLATE-CULTIVATION OF THE SAME NON-LIQUEFYING COMMA-BACILLI OF NOMA AS IN FIG. 32 99

34, 35. SAME NON-LIQUEFYING COMMA-BACILLI GROWING IN 10 PER CENT. NUTRITIVE GELATINE; SEVERAL WEEKS OLD 102

36. CULTIVATION IN 10 PER CENT. NUTRITIVE GELATINE OF MICROCOCCUS ISOLATED FROM THE BLOOD OF THE FINGER OF A PERSON AFFECTED WITH SCARLATINA . 106

37. CULTIVATION IN GELATINE (10 PER CENT.) OF SAME MICROCOCCUS AFTER THREE WEEKS, SHOWING A LARGE FUNNEL-SHAPED OPENING ON SURFACE WITH AN OCCLUDING AIR-BUBBLE 106

LIST OF ILLUSTRATIONS.

FIG.		PAGE
38.	Specimen of Mucus-Flakes from a Monkey.	164
39.	From a Preparation of fresh Mucus-Flakes from the Ileum of a Typical rapidly fatal Case of Cholera	168
40.	From a Preparation of fresh Mucus-Flakes from the Ileum of another Typical rapidly fatal Case of Cholera	169

THE
BACTERIA IN ASIATIC CHOLERA.

INTRODUCTION.

IN the following pages I propose to give an account of the present state of our knowledge of the etiology of Asiatic cholera, gained chiefly in the time that has elapsed since the first communications of Koch on this disease. I do not mean to imply that the observations made by Koch, and others since, are to be regarded as the only valuable addition to our knowledge of this very dire plague; for I am quite aware that all that is of importance in our knowledge of the mode of its spreading and propagation, of the various conditions of soil, air, temperature, water, which affect it—all in fact that has helped us to combat the malady, was gained many years before bacteria and disease-germs had emerged from the region of mystery, long before the recognition of and experimentation with disease-germs had become a branch of exact science. I have only to refer to the works of v. Pettenkofer, of Bryden, Bellew, Cunningham, Budd, Parkes, Snow, and many other experienced observers in India and Europe, to remind the reader, that all that has proved of value from an epidemiological and general sanitary point of view has been gained irrespective of any knowledge of bacteria. The laws governing the spread of cholera are and have been well understood by sanitarians; the fact that cholera, like other infectious diseases, is a communicable disorder spreading from a focus of infection has been long known; and the measures required (and now almost everywhere admitted as necessary) in order to check its spread have been carried out for many years, and without any exact knowledge of diseased-germs. True it was always felt, and a reference to the theoretical parts of the papers by v. Pettenkofer, Bryden, Budd, Parkes, and others fully bears this out, that the knowledge thus gained was of a purely empirical kind, no direct or exact experiment being possible to demonstrate the truth or fallacy of the various measures recommended and employed; and this again was entirely due to the fact that the nature and character of the infective essence of the disease, or the *contagium* (used here in its wider sense), was unknown. While on the one hand there can be no doubt that the discovery of the cholera-germ (which I may at once say has not yet been made) could not in any way alter the nature and application of the general laws of sanitary science, as specially applied to the group of infectious diseases dependent on filth—and according to the general consensus of experienced observers, Asiatic cholera belongs to this group—there can be, on the other hand, no doubt that the identification of a cholera-germ, the knowledge, which from such a discovery would inevitably follow, of its nature and mode of spread, of its mode of alteration by temperature, soil, and season, would unquestionably lead to a more specific application of means to ends than has hitherto been the case. Besides, by an exact knowledge of the cholera-germ, we should be enabled accurately to determine the mode of invasion of the human body by the cholera-virus, the distribution of this latter in the body, the changes it undergoes, the manner in which it leaves the body, and many other very important questions, which would at once emerge from the region of debatable points wherein they at present are; in other words this knowledge would give us a thorough insight into the whole etiology and pathology of the disease, which at present we do not possess. One important series of facts would by the discovery of the cholera-germ at once become plain, viz. the mode of its entrance into the human body and the mode of its exit. At present opinion is divided on both these most essential questions. How does the cholera-virus enter the human body? Does it enter by the alimentary canal only, as is maintained by many authorities, or does it enter also by the respiratory organs? How does it leave the infected body? Is it present in the vomit and discharges from the bowels, as is maintained by most observers; or is it present in these not as an actual but as a potential virus, as is maintained by other equally great authorities? It is obvious that, according to either one of these theories, the mode of our action in combating the spread of the disease ought to become exact and specific.

Now, it is maintained by Koch, and many others who confirm or accept Koch's statements, that in *cholera asiatica* the intestines (chiefly the ileum of the small intestine) of a person affected with the disease is the seat of the rapid growth and multiplication of a definite bacterium (comma-bacillus), which is not present in the blood or any other part of the body, which by its multiplication in the intestine produces a special chemical poison; this poison is absorbed into the system and sets up the whole chain of disturbances of the nervous, vascular, and respiratory organs characterising cholera; and that therefore the bowel discharges containing those specific bacteria are *par excellence* the vehicle of the cholera-germ. It is obvious, that if this comma-bacillus of Koch is in reality the cholera-germ, is really the cholera-bacillus—as truly as the anthrax-bacillus present in the blood of the whole body of an animal affected with or dead from anthrax, and the tubercle-bacillus of Koch present in the tuberculous deposits of an animal or human being affected with tuberculosis, are the germs of these diseases—then the whole series of measures required for checking the spread of cholera would be as simple as they would be efficient. For if it be true that this comma-bacillus, which as we shall see is present only in the contents of, and of course in the discharges from, the bowels of a person affected with cholera, is really and truly the cholera-germ, then the destruction of all the bowel discharges of a cholera-patient, the prevention of bowel discharges gaining access by food, drink, or otherwise to the alimentary canal of others, would be no doubt an effectual and almost the exclusive mode of checking the spread of the disease. Other questions: such as the relation of this comma-bacillus to various conditions of temperature, soil, season, &c., although important from a scientific point of view, would be insignificant compared with this cardinal and fundamental fact. Has it, then, been proved that the comma-bacillus of Koch is the real cholera-germ? This is the question to answer which the following chapters are devoted.

CHAPTER I.

THE HISTORY OF THE COMMA-BACILLUS.

A GOOD many statements of microscopists are on record concerning the occurrence of various forms of bacteria in the dejecta of cholera-patients,[1] but since these statements referred to gross morphological characters only they were not considered of great value. To say that there occurred in the dejecta of cholera-patients micrococci, bacilli, and vibrios, is not one whit more than to say that in human fæcal matter occur these same forms of bacteria. Mr. Fowke[2] claims for Brittan and Swain to have shown in 1849 the occurrence in the choleraic dejecta of the comma-bacilli in the shape of peculiar circular and semicircular corpuscles, which were declared by them not only to be peculiar, but also to have a causal relation to cholera morbus. Looking at the drawings and descriptions reproduced by Mr. Fowke, it does not impress me that these corpuscles are identical, as they are claimed to be, with Koch's comma-bacilli, but I am rather inclined to think that what is there depicted and described are altered and decolourised blood-discs. There

[1] Hassall, Bristowe, Klob, Lewis and Cunningham, and others.
[2] *British Medical Journal*, March 21, 1885.

can be, however, no question about this, that if these corpuscles are really the same as Koch's comma-bacilli, their discoverers did not establish, and by reason of the then crude state of bacteriological research could not have established, that they represented a definite species.

The first account of the presence in the cholera-intestine and cholera-dejecta of a definite species of bacteria characteristic of cholera was given by Koch. We give his own words.[1]

"When we examined the intestine and its contents under the microscope, it was seen that, in some cases, especially in those in which the Peyer's glands were red at the edge, an invasion of bacteria corresponding to this redness had taken place. The bacteria had partly forced their way into the utricular glands, partly pushed themselves between the epithelium and the basement-membrane, thereby lifting the epithelium as it were. In other parts it was seen that they had forced their way deeper into the tissue. Then cases were found in which, behind these bacteria, which had a special appearance with regard to size and shape, so that one could distinguish them from other bacteria and devote special attention to them, various other bacteria forced their way into the utricular glands and the surrounding tissue, *e.g.*, large thick bacilli and very thin bacilli. Thereby conditions were produced similar to those in necrotic diphtheritic changes of the mucous membrane of the intestine and in typhoid ulcers, where afterwards other nonpathogenic bacteria force their way into the tissue rendered necrotic by pathogenic bacteria. We were, therefore, from the very beginning, obliged to look upon these first-mentioned bacteria

---

[1] *Conferenz zur Erörterung der Cholerafrage, Berliner klin. Woch.* 31, 1884. Translated in the *British Med. Journal*, August 30 and September 6, 1884.

as not altogether unimportant for the cholera-process, whilst everything else gave the impression that it was something secondary; for the bacteria first described always advanced beyond the others, they forced their way farther in, and gave one the impression that they had smoothed the way for the other bacilli.

"With regard to the contents of the intestine, at first no clear idea could be formed, as the only cases which came before us for examination were not suitable; in these, also, the contents of the intestine were already putrid and bloody. There were an enormous quantity of various bacteria in these contents, so that there was no possibility of attending to the real cholera-bacilli. Not till I had dissected a couple of acute and uncomplicated cases, in which no hæmorrhage had as yet set in, and in which the contents of the intestines had not yet turned to putrid decomposition, did I recognise that the purer and fresher the cases the more did a special kind of bacteria prevail in the contents of the intestines also, and it was soon clear that these were the same bacteria which I had seen in the mucous membrane. This discovery naturally turned my attention more and more to this kind of bacteria. I investigated them in all kinds of ways in order to establish their special peculiarities; and am able to give the following information regarding them.

"These bacteria, which I have called comma-bacilli on account of their peculiar shape, are smaller than the tubercle-bacilli. One scarcely forms a correct idea of the thickness, length, and breadth of bacteria by giving their dimensions in numbers; I therefore prefer to compare the dimensions of bacteria with other objects, so that one can immediately form a tolerably good idea. As the tubercle-bacilli are known to everybody, I will compare the cholera-bacteria with them. The cholera-bacilli are about half, or at most two-thirds, as long as tubercle-bacilli, but much more bulky, thicker, and slightly curved. This curve is generally not more marked than that of a comma; but sometimes it is larger, becoming semicircular. In other cases it is seen that the curve is doubled, that one comma is attached to another, but in an opposite direction, so that it forms the shape of S. I think that, in both cases, two individual ones after being divided have remained stuck together, and accordingly give the appearance of a more marked curve. But in the artificial cultivations, besides these, another very remarkable form of development of the comma-bacillus is to be found, which is very characteristic of it.

"The comma-bacilli frequently grow in threads of longer or shorter length. But they do not then form straight threads, like other bacilli, for instance, anthrax-bacilli, or, as it appears in the microscopic picture, simple wavy threads, but very long slender spirals, which, as far as their length and the rest of their appearance are concerned, bear the closest resemblance to the spirochætæ of relapsing fever. I could not distinguish one from the other if I had them side by side. Owing to this peculiar form of development, I am also inclined to the view that the comma-bacillus is not a genuine bacillus, but that it is, properly speaking, a transition-form between bacilli and spirilla. Perhaps, indeed, we have here to deal with a genuine spirillum, of which we have a fragment before us. It is seen also in other spirilla—for instance, in spirillum undula—that very short specimens do not form the complete thread of a screw, but only consist of a short little staff, which is more or less curved.

"The comma-bacilli can be cultivated in meat-broth. They grow in this liquid extremely quickly, and in great numbers; and this property of theirs can be utilised for studying their other qualities, by examining with a strong magnifying power a small drop of meat-broth cultivation on a cover-glass. It is seen then that the comma-bacilli move in a very lively manner. When they are collected together at the edge of the drop, and are moving about amongst one another, they look like a swarm of dancing midges, and those long spiral threads appear also moving in an animated manner, so that the whole affords a strange and extremely characteristic picture.

"But the comma-bacilli also grow in other liquids, and especially, in great abundance and speedily, in milk. They do not make milk curdle, and do not precipitate the casein, which many other bacteria, which can also be raised in milk, do. Hence the milk looks quite unchanged; but if you take a small drop from the surface, and examine it under the microscope, it teems with comma-bacilli. They also grow in the serum of blood, in which they also very quickly develop and multiply in great numbers. A very good soil for the reproduction of comma-bacilli is also nutritive gelatine. This gelatine also serves for facilitating and securing the discovery of comma-bacilli; for the colonies of comma-bacilli assume in the gelatine a most characteristic and definite form, which, so far as I can discern, and as far as my experience reaches, no other kind of bacteria assumes in like manner.[1]

"The colony looks, when it is very young, like a very pale and tiny little drop, which is, however, not quite circular, the shape generally assumed by these bacteria-colonies in gelatine; but it has a more or less irregularly bordered, hollowed out, in parts also rough or jagged, shape. It also has, at a very early stage, rather a granulated appearance, and is not of such regular character as in other colonies of bacteria.

[1] This is not strictly correct, as will be shown later.

"When the colony becomes somewhat larger, this granulation becomes more and more evident; at last it looks like a little heap of strongly refracting granules. I might best compare the appearance of such a colony to the appearance of a little heap of pieces of glass. As they grow, the gelatine liquefies in the immediate neighbourhood of the bacteria-colony, and this latter sinks down at the same time deeper into the mass of gelatine. A funnel-shaped cavity is thus formed in the gelatine, in the midst of which the colony is seen as a little whitish point. This appearance is also quite peculiar; it is seen, at least in this manner, in very few other kinds of bacteria, and, as far as I know, never so marked as with the comma-bacilli. The sinking of the colonies can be best observed when carrying out an artificial cultivation. A suitable colony is selected on the gelatine-plate under a microscope with a glass of slight magnifying power; it is touched with a platinum-wire, previously heated; the bacilli are transferred by the wire into a test-tube with gelatine, and this is closed with sterilised wadding. A cultivation of this kind then grows in the same manner as the colony on the gelatine-plate. I am in possession of a numerous collection of artificial cultivations of bacteria made in this manner; but I have never seen in their case such changes as the comma-bacilli cause after being transferred into the gelatine. Here, also, as soon as the cultivation begins to develop you see a little funnel, which marks the point where the inoculation took place. By degrees, the gelatine liquefies in the neighbourhood of this point of inoculation; then the little colony is plainly seen, extending itself more and more; but a deep spot, sunken in, always remains, which looks, in the partially liquefied gelatine, as if an air-bubble were hovering over the colony of bacilli. It almost gives one the impression that the bacillary growth not only causes a liquefaction of the gelatine, but also a speedy evaporation of the liquid formed. We already know a number of other kinds of bacteria which, in quite the same manner, gradually liquefy the gelatine in test-tubes, starting from the point of inoculation. But in these cases there is never such a cavity, nor this bubble-like hollow space.[1] I must further mention that the liquefaction of the gelatine, starting from a single isolated colony (the best way of observing it is in a layer of gelatine, which is spread out on the glass plate), never spreads very wide. The diameter of the liquefied district of a colony may be estimated at one millimetre.[2] Other kinds of bacteria can liquefy the gelatine to a much greater extent, so that a colony attains a size of one centimetre in diameter, and more. In the cultivations of comma-bacilli made in test-tubes, the liquefaction of the gelatine extends by degrees and very slowly, starting from the point of inoculation; and continues in such a manner that, after about a week, the whole contents of the tube have become liquid. Unimportant as all these qualities seem in themselves, special weight is to be laid on them, because they serve to distinguish comma-bacilli from other kinds of bacteria.

"Comma-bacilli can also be cultivated on Ceylon moss (*Agar-agar*), to which meat-broth and peptone are added. This agar-agar jelly is not liquefied by the comma-bacilli. They can also be raised on boiled potatoes—a fact which is very important for certain questions. They grow on potatoes exactly like the bacilli of glanders. The cultivations of comma-bacilli, when grown on potatoes, look like those of glanders-bacilli, but are not coloured so intensely brown, rather a light greyish-brown.

[1] As will be shown later, this is not correct.
[2] This also is not correct.

"Comma-bacilli flourish best at temperatures between 30° and 40° Cent. (86° to 104° Fahr.), but they are not very susceptible to lower temperatures. Experiments have been made on this point, which show that they can grow very well at 17° Cent., though more slowly. Below 17° Cent. the growth is very slight, and seems to cease below 16°. In this point the comma-bacilli remarkably resemble anthrax-bacilli which also have this minimum temperature as the limit for their growth-power. Once I made an experiment to test the influence of lower temperatures on comma-bacilli, and to see if, at a very low temperature, they are not only hindered in their development but also if they cannot possibly be killed. For this purpose an artificial cultivation was exposed for an hour to a temperature of 10° Cent. below zero; during this time it was completely frozen. When part of it was put into the gelatine, there was not the least difference visible in the development or growth, so that they bear frost very well. It is not the same with the withdrawal of air and oxygen. They immediately cease to grow when deprived of air, and accordingly belong, if the division into aërobic and anaërobic bacteria be held as good, to the aërobic class. Any one can convince himself of this very simply, by laying a piece of talc or mica over the glass plate, when the portion of the artifical cultivation has been placed on it in liquid gelatine, and when the gelatine is beginning to stiffen; the talc or mica must be as thin as possible, and must cover at least one-third of the gelatine surface in the middle. The piece of mica, owing to its elasticity, adheres completely to the surface of the gelatine, and thus cuts off the air from the portion covered. Then, as soon as the development of the colonies follows, it is seen that the development only takes place where the gelatine is not covered, and only a

1.] THE HISTORY OF THE COMMA-BACILLUS. 13

trifle, about two millimetres, under the mica plate, up to which point the air has been able to force its way. But under the mica-plate itself nothing grows. Extremely small colonies, invisible to the naked eye, do, it is true, appear, which probably owe their origin to the oxygen existing in the gelatine, but they do not increase in size afterwards. An experiment was made in another manner. Little glasses containing nutritive gelatine, which had been inoculated with comma-bacilli, were placed under an air-pump, and others prepared in the same manner were kept outside the air-pump. It was then seen that those under the air-pump did not grow, but only those outside it. But when those that had been under the air-pump were again placed in the air, they began to grow. Hence they had not died; they only wanted the necessary oxygen to be able to grow. The same occurs when the cultivations are brought into an atmosphere of carbonic acid. Whilst the cultivations that have been kept for comparison outside the carbonic acid atmosphere grow in the usual manner, those that are in a stream of carbonic acid remain undeveloped. But in this case, also, they do not die; for, after having been for some time in the carbonic acid, they begin to grow immediately after they have come out of it.

"On the whole, comma-bacilli, as I have repeatedly observed, grow extremely rapidly. Their vegetation very speedily reaches a maximum, at which it only remains stationary for a short time, then diminishes again very speedily. The comma-bacilli, when wasting away, lose their shape; they appear at one time shrivelled, and at another time swollen, and in this state they are not at all, or only slightly, susceptible to colour. The peculiar conditions of vegetation of comma-bacilli can be best observed by bringing substances which are rich in comma-bacilli, but also contain other bacteria, *e.g.*, the contents of a cholera-intestine or cholera-dejecta, in contact with moist earth, or by spreading them out on linen, and keeping them in a damp condition. Comma-bacilli then increase visibly in a very short time, and in an extraordinary manner in twenty-four hours. Other bacteria that exist with them are at first stifled by the comma-bacilli, a natural pure culture is formed, and, on examining with the microscope the mass that is taken from the surface of the damp earth or linen, preparations can be obtained which show almost exclusively comma-bacilli.

"But this luxuriant growth of comma-bacilli does not last long. After two or three days they begin to die off, and other bacteria then increase. The conditions become the same as in the intestine itself. There also a rapid multiplication takes place; but when the real vegetation-period, which only lasts for a short time, is over, and especially when exudations of blood into the intestine take place, the comma-bacilli disappear, and the other bacteria, especially putrefaction-bacteria, commence to develop in their place. I am, therefore, almost inclined to believe that, if the comma-bacilli were brought at first into a putrefying liquid which contained a great deal of the products of vital changes of other bacteria, and especially of putrefaction-bacteria, they would not come to development, but would soon die off. But so far sufficient experiments have not been made on this point; it is only a supposition which I make, supported by my experiences of other bacteria-cultivations. This point is important, because it is not a matter of indifference whether the comma-bacilli, if they come into a sink or sewer, find a good or a very bad soil for reproduction. In the first case, they would multiply, and would have to be destroyed by methods of disinfection;

but in the latter case they would die off, and there would be no necessity for disinfecting. I am inclined to hold the latter view, as borne out by all the experience I have so far had.

"In these cultivation-experiments it was further seen that the nutritive substances—at least, the gelatine and meat-broth—must not be acid. As soon as the gelatine shows only a trace of acid reaction, the growth of the comma-bacilli is very stunted. If the reaction be in a marked degree acid, the development of the bacilli completely ceases. It is at the same time noteworthy that it is not all acids that seem to be unfavourable to the comma-bacillus; for the surface of a boiled potato, where it is cut, is known to have an acid reaction, in consequence, if, I am not mistaken, of its containing malic acid. Nevertheless, comma-bacilli grow very luxuriantly on potatoes. Hence, one cannot say, straight off, that all acids hinder the growth; but, in any case, there are a number of acids which have this effect. In meat-broth it is probably lactic acid, or an acid phosphate.

"In these experiments on the influence of substances in arresting the development of comma-bacilli, the striking fact was evident that comma-bacilli die off extremely easily when dried. These experiments were made by letting a very small drop of a substance containing bacilli dry on a cover-glass, and a large supply of these cover-glasses was immediately prepared for a series of experiments. A drop of the liquid which was to be examined was then placed upon such a cover-glass, and left for development in the hollow glass slide. Having proceeded in this manner, in no single preparation did anything grow that had received meat-broth as nutritive fluid, nor in a striking manner in the test-preparations either. At first I did not know what caused the absence of growth, and thought that the broth must be the cause of it, for I have never met with anything like this before in the case of other bacteria. For instance, anthrax-bacilli can be kept in store for a long time dry on cover-glasses; they retain vitality from half a week to nearly a whole week in this manner. As, however, the meat-broth on examination proved to be unexceptionable, we had to examine whether the comma-bacilli had not probably died off owing to being dried upon the cover-glass. In order to obtain certainty on this point, the following experiment was made. A number of cover-glasses were provided with a small drop of substance containing bacilli. The drop dried up in a few minutes. One cover-glass was now charged with a drop of meat-broth after an interval of a quarter of an hour, another after an interval of half an hour, another after an interval of an hour, and so on. Then it was seen (and I made several series of experiments) that the comma-bacilli did come to development on the dried glass plates that had lain a quarter, a half, and a whole hour, but after two hours they sometimes died off; after three hours, I could not keep the bacilli alive in these experiments. Only when compact masses of bacilli-cultivations—for instance, when the pulpy substance of a cultivation made on potatoes was dried—did the bacilli retain vitality for a longer time; clearly because in this case complete desiccation followed much later. But even under these conditions I have never succeeded in preserving the bacilli alive in a dried state longer than twenty-four hours.

"This result was so far important, that by its means it could easily be tested whether the bacteria have a resting state. We know that other pathogenic bacteria—for example, anthrax-bacteria, which form spores—can be preserved for years in a dry state on a cover-glass without their dying. We know also of other infectious substances with whose nature we are not yet accurately acquainted, for example, the infectious matters of small-pox and of vaccine, which can be kept in a dried state for several years, still retaining their power of infection. If now the comma-bacilli, which, as such, are very speedily killed by drying, pass into a resting condition under some circumstances, that would very soon be shown during the process of drying.

"This is always one of the most important questions in the etiology of an infectious disease, and especially so of cholera. The investigation of this point has therefore been made in the most careful manner possible, and by every possible method, and I hardly think that anything more can be done on this point. Above all, cholera-dejecta and the contents of the intestines of cholera-corpses were left in a damp condition on linen, in order that the comma-bacilli might develop under the most favourable circumstances. After certain intervals of time, pieces of the linen were dried—for example, after twenty-four hours, after a few days, after several weeks—to see if during this period any condition of permanence had been established. For infection through cholera-linen affords the only undisputed example of the presence of an effectual infectious substance which adheres to a special object. If there were a permanent or resting state to be found anywhere, it must have been found on cholera-linen.

"But in none of these cases was a permanent state discovered. When the dried objects were examined, it was seen that the comma-bacilli had died off. Then, further, the dejecta were placed in earth, being either mixed with earth or spread on the surface, which was either kept dry or moist; they were mixed with marsh-water; and were also left to decay without anything being added to them. In gelatine-cultivations, the comma-bacilli have been cultivated up to six weeks, also in serum of blood, in milk, and on potatoes, on which anthrax-bacilli are known to form spores extremely rapidly and in great abundance. But we have never obtained a permanent state of the comma-bacilli. As we know that the majority of bacilli have a permanent or resting state, this result appears very striking. But I will remind you here of what I mentioned before, that we have most probably to deal with a micro-organism which is not a genuine bacillus at all, but is more allied to the group of screw-shaped bacteria, or spirilla; but we do not know of any permanent state of spirilla as yet. Spirilla are bacteria which depend for their existence exclusively on liquids, and do not, like anthrax-bacilli, vegetate under certain conditions in which they have for once to endure a dry state. It therefore seems to me, as far at least as my experience goes, that there is no prospect of finding a permanent state of comma-bacilli.

"In accordance with the cholera-material that I have so far examined, I think I can now assert that *comma-bacilli are never found absent in cases of cholera;* they are something that is specific to cholera.

"As a test, a considerable number of other corpses, dejecta from patients and persons in good health, and other substances containing bacteria, were examined to see if these bacilli, which were never missing in cases of cholera, might, perhaps, occur elsewhere also. This is a point of the greatest importance in judging of the causal connection between comma-bacilli and cholera.

"Amongst these objects for investigation was the corpse of a man who had had cholera six weeks before, and had afterwards died of anæmia. There was no farther trace of

I.] THE HISTORY OF THE COMMA-BACILLUS. 19

comma-bacilli to be found in his intestines. The dejecta of a man who had had an attack of cholera for eight days previously were also examined; his stools were already beginning to be consistent; in this case also comma-bacilli were absent.

"I have also thoroughly examined more than thirty corpses, in order to convince myself more and more that these bacilli are really only found in cases of cholera. Corpses of those who had died of affections of the intestines, *e.g.*, of dysentery or of those catarrhs of the intestine frequently mortal in the tropics, were chiefly selected for this purpose; also cases with ulceration in the intestine, a case of enteric fever, and several cases of bilious typhoid.

"In the last-named disease, the modifications in the intestines are at first sight very similar to those which take place in severe cases of cholera, in which hæmorrhage of the intestine takes place. The small intestine is in its lower section infiltrated by hæmorrhage; but, strange to say, this change in bilious typhoid affects mostly the Peyer's patches, whilst in cholera these are very little changed.

"In all these cases, where we had to deal chiefly with diseases of the intestine, no trace of comma-bacilli was to be found. Experience teaches that such affections of the intestine make people especially liable to cholera. So one might have pre-supposed that comma-bacilli, if they were to be found anywhere else, must be found in these cases. Besides these, dejecta of a large number of dysenteric patients were examined without the comma-bacilli ever being met with. I continued these investigations afterwards in Berlin, together with Dr. Stahl, my untiring fellow-labourer, a man who promised much for the investigation of bacteria, had not an early death unhappily put an end to his work. We examined a considerable number of various dejecta, especially of children's diarrhœa, as well as that of grown-up persons; saliva also, and the mucus that adheres to the teeth and tongue, which abounds in bacteria, for the purpose of finding comma-bacilli, but always without success. Various animals were also examined with this view. Because a complication of symptoms very similar to those of cholera can be obtained by arsenical poisoning, animals were poisoned with arsenic, and afterwards examined. A great number of bacteria were found in the intestines, but no comma-bacilli. Nor were they found in the sewage from the drains of the town of Calcutta, in the extremely polluted water of the River Hooghly, in a number of tanks which lie in the villages and between the huts of the natives and contain very dirty water. Everywhere, where I was able to come across a liquid containing bacteria, I examined it in search of comma-bacilli, but never found them in it. Only once did I come across a kind of bacterium which, at first sight, bore a strong resemblance to comma-bacilli, and that was in the water which, at high-tide, floods the margin of the salt-water lake that lies to the east of Calcutta; but, on a closer inspection, they appeared larger and thicker than comma-bacilli, and their cultivations did not liquefy gelatine.

"Besides these observations, I have had a considerable experience in bacteria, but I cannot remember ever having seen bacteria resembling the comma-bacilli. I have spoken to several people who have made a great number of cultivations of bacteria, and have also had experience, but all have told me that they have not as yet seen such bacteria. I therefore think I may say positively that the *comma-bacilli are constant concomitants of the cholera-process*, and that *they are never found elsewhere.*"

These statements are very definite and precise, and the

description give by Koch of the distribution, morphological characters, and cultivation of the comma-bacilli are very detailed and clear, and it cannot be said that any other observer has been able to add anything of importance since Koch's publications. The statements by Von Ermengem, Babes, Watson Cheyne, and others on the morphological and culture characters of the comma-bacilli, are therefore to be regarded as repetitions of those first made by Koch. Nothing new is brought forward by these observers. I think I may therefore be excused from referring to the statements of these authors so far as they treat of this part of the subject.

CHAPTER II.

THE DISTRIBUTION OF COMMA-BACILLI.

IN the preceding account by Koch we see, then, that (*a*) the comma-bacilli occur in the intestinal dejecta during the acute stage of the disease : (*b*) the comma-bacilli are present in the mucus-flakes and in the fluid of the contents of the small intestine, chiefly and most numerously in the lower portion of the ileum ; towards the upper part of the ileum their number decreases, and in the jejunum they become very scarce—hence the vomit is as a rule free from them, and when they are present it is no doubt owing to regurgitation : (*c*) the comma-bacilli are present in the tissue of the mucous membrane of the lower ileum, in the epithelium of the surface, in the lymphatic tissue of the mucosa, within the cavity of the crypts of Lieberkühn as well as between the epithelium lining these crypts and their limiting membrana propria, but especially in the lymph-follicles of the Peyer's patches ; these according to Koch are in pure acute typical cases visible as swollen hyperæmic structures, the blood-vessels of the marginal portion being distended and filled with blood, and hence strongly marked ; in these blood-vessels Koch states that he found them in great abundance : (*d*) the more acute and typical a case of cholera, the more numerous are the comma-bacilli found in the lower ileum, so much so that in the very acute cases, marked by the whole chain of symptoms characteristic of a typical case of cholera, the lower ileum contains the comma-bacilli "almost in pure cultivation": (*e*) no comma-bacilli occur in the blood of the general circulation, in the mesenteric glands, or any other organ.

The observations which I have made with regard to the general distribution of the comma-bacilli enable me to say this, that while agreeing with Koch in some, I differ from him in other very essential, points. We shall take the above statements seriatim.

(*a*) There can be no question about this important fact, that in every case of acute cholera during the first days, *i.e.* while the patient suffers from severe purging, the intestinal discharges contain the comma-bacilli; but there does not exist, according to my experience, extending over a considerable number of cases, any definite relation between the number of comma-bacilli present in the stools and the severity of the disease. I have examined a good many stools of patients during the first day or first two days of illness, all the symptoms characteristic of typical cholera being present —severe vomiting and purging of watery fluid containing mucus-flakes, great fall of temperature, voice and face that of cholera, suppression of urine, respiration very irregular and oppressed—and yet the most careful examination of fresh preparations and of preparations stained in the usual manner revealed a few comma-bacilli only. In one instance only have I come across a stool containing very numerous bacilli. This stool was almost clear watery fluid in which were suspended minute greyish flakes; under the microscope a great many comma-bacilli were found, and but few other bacteria, the small mucus-flakes being almost like a pure cultivation of the comma-bacilli. I must however state here, that stools more or less fluid and of a fæcal character (*i.e.* not simply watery fluid containing mucus-flakes in suspension, but with finely distributed particles of fæcal matter) never contained comma-bacilli in conspicuously large numbers. They were present amongst crowds of other bacteria, either as isolated, slightly curved commas, as semicircular or circular corpuscles, some conspicuous by their small size, as S-shaped or dumb-bell-shaped particles, and as shorter or longer spirals. Only a few groups of them were present.

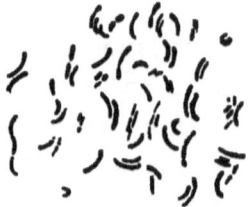

Fig. 1.—From a Preparation of Fresh Mucus-Flakes from a Choleraic Evacuation.

Showing large numbers of comma-bacilli and a good many minute straight bacilli. Amongst the comma-bacilli there are a few small semicircular ones. Magnifying power about 700.

(*b*) A much better insight into the distribution of the comma-bacilli is obtained by examining the intestinal contents taken directly from the small intestine, which of course can only be done on making the post-mortem examination. In some acute typical cases on opening the abdominal cavity the small intestine appears much congested and distended, and in its interior is present a grumous fluid not large in quantity and of a brownish colour, containing amongst particles of fæcal matter mucus-flakes and small clots of blood. In these cases the mucosa is streaked and dotted with blood. In other cases of rapid death the whole of the small intestine, including the upper part of the jejunum and duodenum, is of a rosy tint; here and there the colour is in patches and more pronounced, the intestine is distended, and on opening it a large amount of watery fluid escapes in which are suspended very numerous flakes of various sizes. The internal or mucous surface is pale, and the epithelial layer is in many places loose or separating in larger or smaller flakes. Placing a piece of the intestine under water this loosened or detached condition of the epithelium becomes very conspicuous. In the lower ileum rarely and

Fig. 2.—Preparation of Mucus-Flakes from the Ileum of an Acute Case of Cholera.

1. Masses of single comma-bacilli.
2. Circular forms.
3. Semicircular forms.

Magnifying power about 1,400.

then only in a few places are present small particles of fæcal matter adhering to the mucous membrane, but as a rule only watery fluid and mucus-flakes are present. Such cases correspond to Koch's "pure" cases; the symptoms during life are always vehement and well pronounced, and death ensues generally during the first twenty-four hours. There is much purging with rice-water stools. In such cases, according to Koch, the lower ileum is almost a pure cultivation of comma-bacilli. But I cannot confirm this statement.

I have seen several such pure cases of acute cholera in which the mucus-flakes suspended in the fluid of the cavity of the small intestine, and the epithelial flakes loosened but still adhering to the mucous membrane, were very carefully examined in stained specimens, but there was rarely such a condition as an almost pure cultivation of comma-bacilli. There were present in the fluid and in the flakes crowds of bacteria; in some instances and in some flakes the comma-bacilli were extraordinarily numerous, and almost in a state of purity, in others they were scarce—and in fact there were

Fig. 3.—Preparation of Mucus-Flakes from the Ileum of an acute Case of Cholera.
1. Bacterium termo, probably v. Emmerich's bacillus.
2. Comma-bacilli.
3. Minute straight bacilli.
Magnifying power about 1,400.

some cases where there was difficulty in finding them. What always appeared to me a curious point and difficult to understand was this: in several post-mortem examinations of pure acute cases the anatomical characters of the whole of the small intestine, jejunum and ileum, and its contents were very much the same, yet on microscopic examination of the fluid and the mucus-flakes it was generally found that the number of bacteria present in the jejunum were very small, and gradually increased towards the lower ileum; and this held good not only for the comma-bacilli but also for the other bacteria present. In addition to this a comparison of the different cases which came under my observation showed this significant fact, that in many instances in which the post-mortem examination was delayed, the number

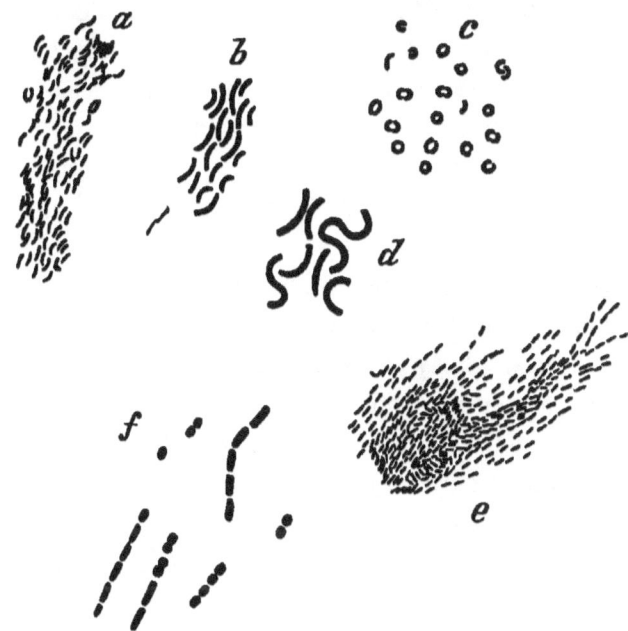

Fig. 4.—From a Preparation of Fresh Mucus-Flakes from the Lower Part of the Ileum of a Typical rapidly Fatal Case of Cholera. (Duration of illness nine hours and a half, post-mortem examination after one hour.)

    The forms here delineated are met with in the same mucus-flakes.
    a. Masses of minute comma-bacilli.
    b. Masses of typical choleraic comma-bacilli.
    c. Minute circular and semicircular comma-bacilli.
    d. Large thick comma-bacilli.
    e. Masses of the minute straight bacilli.
    f. Micrococcus and thick straight bacilli.
        Magnifying power about 700.

of comma-bacilli and other bacteria was likely to be greater than when the examination was made immediately or almost immediately after death.

In the following (copied from the Report on Cholera by the English Cholera Commission: *An Enquiry into the Etiology of Asiatic Cholera*) is given a tabular statement of the occurrence of bacteria in the mucus-flakes taken from the lower part of the ileum of typical rapidly fatal cases, the ileum being slightly reddened and filled with clear fluid in which were numerous typical flakes. The numbers attached to the cases indicate the number in the total series of cholera cases examined in Bombay and Calcutta.

1. Case 2.—Death after 40 hours. Post-mortem made after four hours. Comma-bacilli abundant, small and large straight bacilli.

2. Case 11.—Death after 18 hours. P.m. after half an hour. Comma-bacilli tolerably numerous; they vary in length, and particularly in *thickness*. Large straight bacilli exceedingly numerous; minute straight bacilli.

3. Case 14.—Death after 12 hours. P.m. after half an hour. Comma-bacilli very scarce. Few other bacteria.

4. Case 16.—Death after 18 hours. P.m. after three-quarters of an hour. Very few comma-bacilli. Exceedingly numerous small straight bacilli, singly and in clumps. Other kinds of bacteria.

5. Case 23.—Death after 20 hours. P.m. after one and a half hours. Various species of bacteria; micrococcus, bacterium termo. Very few comma-bacilli; they are distinctly thinner than those of other cases. Minute straight bacilli in clumps.

6. Case 32.—Death after 27 hours. P.m. after two hours. All kinds of straight bacilli in great numbers. The small straight bacilli numerous. Comma-bacilli tolerably numerous; they are of different lengths and *thicknesses*.

7. Case 35.—Death after 13 hours. P.m. after a quarter of an hour. Comma-bacilli tolerably numerous; large

DISTRIBUTION OF COMMA-BACILLI.

straight bacilli tolerably numerous. The small straight bacilli exceedingly numerous.

8. Case 48.—Death after 14 hours. P.m. after half an hour; great abundance of comma-bacilli, and also numerous minute straight bacilli.

9. Case 51.—Death after $9\frac{1}{2}$ hours. P.m. after one hour; various kinds of bacilli. The minute straight bacilli in extraordinary numbers. Comma-bacilli of three different kinds distinguished by their various thicknesses, some exceedingly minute, others five and six times as big, and a third variety corresponding in length and thickness to the typical comma-bacilli of other cases. The first variety in very large numbers, forming continuous masses. Numerous small semicircular commas, corresponding in size to the small variety of the above commas.

All these organisms were numerous in the free flakes, as well as in those still on the mucous membrane.

Other cases, which were typical and rapidly fatal, but in which the ileum did not contain the clear watery fluid with mucus-flakes, are not included here.

Drs. Weisser and Frank ascribe to me in the *Archiv. f. Hygiene* iii. 1. p. 380 the assertion: "That in very rapid cholera cases the comma-bacilli are missed." That this is not my assertion I have stated above; where these gentlemen got hold of it I cannot say, unless, like Koch, they got their information about my statements at second hand. Examining the tables published by Weisser and Frank (*l.c.* pp. 382–389)—in which they give an account of the examination of cover-glass specimens sent them by Dr. Dissent of Calcutta, and made of the contents of the intestine in numerous cases dead of typical cholera—it will be seen that out of thirty-one cases dead within twenty-four hours (seven to twenty-four hours), in fourteen the comma-bacilli were scarce (*wenige* or *spärliche*), in one none could be found; then out of twenty-four cases dead between twenty-four and forty-eight hours there was one without the comma-bacilli, one was questionable, and in nine cases the comma-bacilli were scarce (*spärlich* or *vereinzelt*). These facts then of Drs. Weisser and Frank do not seem to agree with the conclusions they draw (*l.c.* p. 390), but they singularly harmonise with my own statements, viz., that there does not exist any definite and uniform relations between the severity and rapidity of the disease and the number of comma-bacilli present in the intestine, as was maintained by Koch.

(*c*) The statement of Koch that the comma-bacilli in the acute stages are present in the tissue of the mucous membrane of the ileum requires the most serious consideration. If it were true that in the acute stages of the disease, the comma-bacilli are constantly present in the tissue of the mucous membrane in the definite manner described and figured by Koch, *i.e.* in the epithelium and superficial mucosa, around the Lieberkühn's follicles, and in the peripheral zones of the lymph-follicles of the Peyer's glands, then a very important point in the chain of evidence would thereby be established. One of the most essential and generally acknowledged requirements in proving the connection between a definite species of bacterium and the causation of an infectious disease is the constant presence of this particular species in the diseased tissues. Although complete proof is not thereby given, yet it must be obvious that the constant presence in large numbers of a definite species in the diseased tissues cannot be of an indifferent nature. In all those cases of infectious disease in which a definite species of bacterium has been unequivocally proved to be the cause of the disease, this constant presence of that definite species of bacterium has been established.

DISTRIBUTION OF COMMA-BACILLI.

Take anthrax and relapsing fever, glanders and tuberculosis, erysipelas and leprosy. In the first two the blood and spleen are the seat of the *bacilli anthracis* or of the *spirilla Obermeieri* respectively, in the second two the morbid deposits contain the *bacilli* of *glanders* or *bacilli tuberculosis* respectively, in erysipelas the lymphatics at the mar-

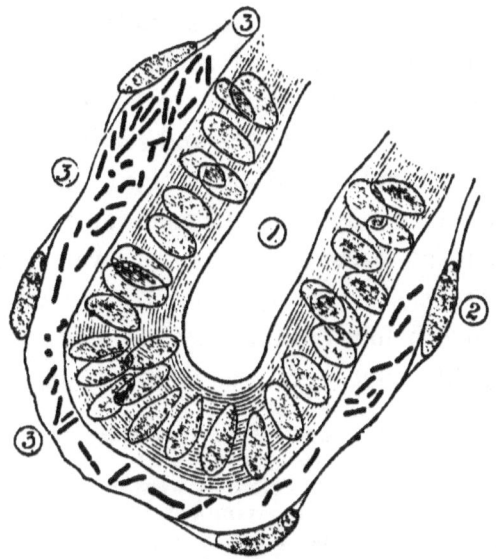

FIG. 5.—FROM A SECTION THROUGH THE ILEUM OF AN ACUTE CASE OF CHOLERA.
In this case no comma-bacilli were present anywhere in the tissue of the mucous membrane.
1. Cavity of a crypt of Lieberkühn, lined with columnar epithelium.
2. Nuclei of the membrana propria.
3. Space between the detached epithelium and the membrana propria; in it numerous straight bacilli.
Magnifying power about 1,400.

gin of the inflamed skin contain the specific *micrococci*, and in leprosy the cells and tissue of the leprous nodules and the lymph-spaces around them contain the *bacilli lepræ*. And the same holds good for other infectious diseases, vaccinia, variola, swine-fever, swine-erysipelas, septicæmia, pyæmic abscesses, &c. &c. If then in cholera the tissue of the intestine, say the ileum, *constained constantly in the acute stages of the disease numbers of comma-bacilli*, I should consider this as of fundamental importance, and I should go so far as to say that one of the most important links of the chain of proof that they are the cause of the disease had been established. I have therefore paid particular attention to this point, and from a very careful examination of an enormous number of preparations made of the ileum and other parts of the intestine of acute and typical cases of cholera, I am able most positively to assert that nothing of the sort occurs[1]. Fresh sections and sections of the tissues hardened in alcohol or Müller's fluid, stained with the different aniline dyes by the usual methods employed for the demonstration of bacteria in tissues, were prepared and examined; the cases were typical and atypical; some were cases that died before the first day was over, all the symptoms during life were very characteristic, on post-mortem

[1] Koch, in his second paper, "Further Researches on Cholera," (*Zweite Conferenz zur Erörterung der Cholerafrage*, Berlin, May, 1885) says of myself:—". . . Even before he went to India his judgment of my statements was formed. He attempted, at that time, to show that I had contradicted myself; that I had, in Egypt, compared the bacteria found in the wall of the small intestine with the bacilli of glanders, but that the latter were not curved, but straight bacilli; then all at once, in India, the straight bacilli had become curved ones." What may have prompted Koch to write this I am unable to say, but this much I can positively say, that at no time or place, neither before I went out to India, nor in India, nor since my return, have I said or written anything of the sort. Koch has evidently been misinformed, and after this I am under the impression that Koch has entered on a criticism of my work without having read what I said. I am sorry to think that he ascribes to me anything so absurd; for it would no doubt be absurd on my part to try to make out that the glanders-bacilli were not curved after having myself figured them as curved (see my *Micro-organisms and Disease*, Fig. 62, third edition, 1886); and it would be equally absurd and incorrect on my part to say that Koch had stated while in Egypt that the cholera-bacilli were straight, but that all at once, in India, they became curved. I am sure such criticisms would not have been applied by Koch to myself if he had read what I did say.

examination the condition of the intestine was such as would merit Koch's term of pure acute cholera; the post-mortem was made as soon as possible after death, in some instances as early as a quarter of an hour, in others one and a half to two hours, but I did not find anything that showed the presence of comma-bacilli in the intestinal mucous membrane, not even in the superficial epithelium where this had kept its position. Such appearances as are described and figured by Koch,[1] or anything approaching them, were not met with in one single instance. But there were cases under observation in which the tissue did contain a few comma-bacilli besides other bacteria, and I will describe these here more in detail.

In two cases only were there present in sections through the Peyer's glands near the ileo-cæcal valve comma-bacilli in some places around Lieberkühn's crypts, and also scattered here and there amongst the superficial parts of the lymph-follicles. But besides the comma-bacilli, and in greater numbers, were straight bacilli, which with the comma-bacilli could be traced from the broken surface into the depth of the mucosa. As one of these is a good example of comma-bacilli being found in the mucosa, but accompanied by a larger number of straight bacilli, we will give the history of this case. The patient, æt. thirty, was attacked with vomiting and purging at 4.30 p.m. on the 6th October, he was admitted into the J. J. Hospital, Bombay, at 7.30 p.m. on 7th October. When admitted he was deeply collapsed, pulse imperceptible, features sunken, extremities cold, no urine. He died at 6 a.m. on 8th October. Post-mortem at 8.30 a.m. The patient was evidently moribund from 7.30 p.m. of 7th October till 6 a.m. of 8th October, *i.e.* for

[1] *Loc. cit.* p. 6, fig. 1.

nearly twelve hours; in addition to this the post-mortem was made two hours and a half after death; the temperature of the air was above 75° F. No wonder that under all these circumstances the tissue of the bowels should have become invaded by micro-organisms. In another case of acute typical cholera, where the post-mortem had been made fourteen minutes after death, but where the patient had been moribund from 9 a.m. till 3 p.m., sections through the hardened Peyer's glands and mucosa of the ileum showed the epithelium of the surface as well as that lining the Lieberkühn's follicles bodily loosened and raised from the mucosa, but fixed in position during hardening. While there was total absence of comma-bacilli here or anywhere else in the mucous membrane and lymph-follicles, there were nevertheless in some places on the surface minute groups of putrefactive *bacillus subtilis*, and from here they could be traced into the spaces resulting from the detachment of the epithelium of the Lieberkühn's follicles from the membrana propria. And even capillary blood-vessels of the lymph-follicles near the denuded surface were found crowded with putrefactive bacilli and micrococci. In a third typical case (death after ten hours, post-mortem after half an hour), there were present numbers of straight putrefactive bacilli in the tissue of the villi and around the bottom of the Lieberkühn's follicles, but only here and there could a comma-bacillus be found close to the epithelium of the surface.

From this then we conclude that comma-bacilli as well as other bacteria can find entrance into the tissue of the intestine, but that this in a measure depends on the state of disorganisation of the intestine, and the time that elapses between the stage of "agony" and actual death. That the comma-bacilli take the lead in penetrating the tissue, both as regards depth and number, as is maintained by Koch in regard to the acute stage of cholera, is not borne out even in those cases that are particularly favourable to the immigration of bacteria from the surface into the depth of the tissue.

To say then as Babes does,[1] that in the material (intestine) from a case of cholera, which had been preserved in alcohol for some years, he found comma-bacilli in the tissue, means nothing whatever, since it is not stated how long after death the intestine had been left before being removed to the hardening fluid. The same negative value is to be attached to the statement of Mr. Watson Cheyne, who says that after repeated examinations he found the comma-bacilli in the tissue, though he at first missed them. As has been mentioned just now, I have myself examined such preparations, but there was always clear evidence that the tissue was invaded also and more numerously by other bacteria, or that the tissue in which the comma-bacilli were present was near the surface, and in a state of necrosis or profound alteration. Judging from the numerous examinations of sections of cholera-intestine that have been made by various observers during the last few years, there has not been one confirming their usefulness for the purpose of diagnosis. Klebs, Van Ermengem, Von Emmerich, Buchner and others, have all questioned the statement of Koch.

A point of importance in interpreting the occurrence of bacteria in the diseased intestine is, that it is necessary to bear in mind that bacteria can penetrate during life into the tissue of even a perfectly healthy intestine. Bizzozero was the first to show[2] that in the tissue of the lymph-follicles constituting the Peyer's glands in the rabbit there occur,

---

[1] *Virchow's Archiv*, 1884.
[2] *Centralblatt f. d. med. Wiss.* 1885.

even during life, and in the perfectly normal state, bacteria which can be proved to have penetrated there from the free surface. These bacteria are present between the young cells or enclosed in large lymph-cells, these playing probably the part of scavenger-cells (or phagocytes of Metchnikoff), inasmuch as they swallow and destroy the bacteria that had penetrated into the lymphatic tissue. I have repeated these experiments of Bizzozero, and I can fully confirm his observations. The very first rabbit examined with this object yielded positive results. A perfectly normal and healthy rabbit is killed by decapitation, the abdomen opened, with clean and sterile forceps the serous coat and outer muscular coat are gradually stripped off the so-called vermiform process, which is in reality the beginning of the cæcum; this as is well known is really one continuous mass of lymph-follicles. Having exposed the deepest part of the mucous membrane containing the lymph-follicles, with the sterilised blade of a scalpel a scraping is taken of the deep part of the lymphatic tissue, and with this scraping cover-glass specimens are made, dried, and stained in gentian-violet. On microscopic examination small bacilli are found scattered and isolated between the nuclei of the lymph-cells, and here and there a huge lymph-cell—five to eight times the size of an ordinary lymph-cell—is met with, the protoplasm of which is crowded with the same small bacilli.

From this it follows then, that while in the perfectly normal state bacteria can make their way from the free surface, or internal cavity abounding with bacteria, into the tissue of the mucous membrane, their penetration will be no doubt considerably facilitated if the wall of the intestine is in a state of disease and disorganisation; for, as is well known, living and healthy tissues do not favour, but on the contrary are inimical to, the existence of septic bacteria.

An important line of research is hereby opened up, namely to enquire to what extent in the human subject do the bacteria normally present in the cavity of the intestine penetrate into the wall of the intestine? It must be clear from the above that statements as to the occurrence of one or other species of bacteria in the tissue of a diseased intestine, *e.g.* in typhoid fever, cannot claim that significance which has hitherto been attributed to them. In typhoid fever the disorganisation of the intestinal wall is very profound, and lasts days and weeks; there is no reason why bacteria, particularly motile bacilli, such as the so-called typhoid bacilli are, should not penetrate deep into the intestinal wall and thence into the mesenteric glands, particularly into the foci of inflammation and necrosis always present in these glands in typhoid fever, and even further into the necrotic foci of the spleen. And the same applies to the choleraic intestine; in some cases, particularly those remaining for some hours *in articulo mortis*, or kept for some time after death before examination is made, the motile comma-bacilli and other motile bacilli of the internal cavity can penetrate into the tissue of the intestine, particularly as the mucous membrane is in a profound state of disorganisation, and, as has been noticed and described by various observers, also into more distant localities, *e.g.* the liver and gall-bladder.

(*d*) As I have mentioned under (*b*), I cannot confirm the statement of Koch—that the purer and the more typical and acute a case of cholera, the more does the lower part of the ileum contain an almost pure cultivation of the comma-bacilli. Although I have found in some typical acute cases that the mucus-flakes of the contents of the lower ileum contained comma-bacilli in large numbers and continuous masses, I have seen others where these comma-bacilli were few and far between; and besides whenever the comma-bacilli were very abundant various other forms of bacteria were also present in great numbers. I refer to the list of cases given on a former page (p. 28).

(*e*) I can fully confirm Koch's statement that no comma-bacilli or other bacteria occur in the blood and in the viscera. The blood was examined fresh from living patients in various stages of the disease, in some the disease had not lasted more than six hours, in others sixteen hours, in others twenty-four hours and more. But there were no bacteria of any known forms present in the blood.

Sections of the kidney, liver, spleen, mesenteric glands, made fresh and after hardening, and stained with the usual dyes, were made in large numbers, but no organisms of any known characters were met with. In the gastric discharges (vomit) of cholera patients the comma-bacilli are rarely present, as Koch has already shown; I have myself examined six cases, but have not found them more than once, and then only in very small numbers indeed.

As a result of this part of the subject we find then (1) that comma-bacilli are constantly present in the intestinal contents in acute cases of cholera;[1] (2) their number cannot be said to have any definite relation to the acuteness and severity of the illness, since in some typical and acute

---

[1] Koch charges me in his paper read at the Second Conference on Cholera, held in Berlin on May 4, 1885, with having had to admit that true cholera-bacilli occur in all cases of cholera. From this it might be inferred, as is also definitely stated by Drs. Weisser and Frank (see above), that I at first did not admit such a fact, but finally had to admit it. This is another instance of the manner in which Koch criticises those who differ from him, for I am not aware of having ever said that the choleraic comma-bacilli do not occur in all cases of cholera; there was therefore no occasion for me to alter my statement.

and pure cases they were present only in small numbers together with multitudes of other bacteria; (3) there are no comma-bacilli or other bacteria present as a rule in the tissue of the intestine of acute typical cases of cholera, although in some cases they may penetrate from the cavity into the wall of the intestine.

CHAPTER III.

MORPHOLOGY OF THE CHOLERAIC COMMA-BACILLI.

A.—*The Comma-bacilli of Cholera-stools and of the Intestinal Contents.*

As has been described by Koch, the single comma-bacillus is a minute rod more or less curved, being a portion of a small or large circle. Owing to this shape Koch named it the comma-bacillus, a name which I think unfortunate and inappropriate. As Koch has shown, and as we shall see more fully below, there can be no doubt about the single comma-bacillus being in its full development an element of a vibrio or spirillum, and for this reason it is not appropriate to speak of the species as *Bacillus;* and for the same reason the name cholera-bacillus is not more acceptable; it implies, besides, that this form of bacterium is peculiar to cholera; now although, as will be shown below, this peculiar shape of the organism was unquestionably in the mind of Koch (see *loc. cit.* p. 25) when he described his observations in Egypt, India, and France, it is now known that these so-shaped organisms, *i.e.* comma-bacilli, occur under several conditions other than cholera.

The comma-bacillus, then, occurs in the intestinal discharges and in the contents of the lower ileum, chiefly as single and S-shaped rods, the single rods varying in length between 0·6 and 1·2 μ: their thickness is about 0·2 μ. The greatest differences exist as regards the amount of curvature. While some, particularly the short examples, show only a slight curve, noticeable more at the ends, there are close by in the same particle of the mucus-flake numbers of others which are unmistakably curved. Then one always finds in the same flake, particularly in cases in which the comma-bacilli are abundant, numbers of comma-bacilli in clumps or in streaks, where the majority of the elements show only just an indication of curvature, while others show a distinct curve, some being curved as much as one-half to two-thirds of a circle. Several such instances were examined, in which, in the mucus-flakes, large numbers of semicircular and still more curved elements were present. As regards the mucus-flakes, this can be stated with certainty—that the more rapidly the multiplication of the comma-bacilli proceeds, that is to say, the more they occur in groups, patches, or streaks, the less pronounced is the curvature in the single elements. And the same may be said as to the length of the single elements: the more rapid their multiplication in the flakes, the shorter are the majority of the elements.

Great differences occur also in the thickness of the comma-bacilli in the intestinal contents. While the typical comma-bacilli are about the thickness of 0·2 μ, there are always present some that are twice and thrice as thick, and there can be no question about this fact, that in many specimens taken from the intestinal contents of typical acute cases, there occur comma-bacilli differing from one another in length and thickness within such limits that one might well doubt their belonging to the same species. I refer in this respect to the case 51 described on a former page and to Fig. 4. Here the individuals constituting the groups and masses of short, thin, very slightly-curved comma-bacilli represented at *a* are conspicuously different from the typical well-curved ones represented at *b* and *c*.

Very interesting forms of the comma-bacilli are those in which the curvature amounts to half or two-thirds of a circle, or almost a whole circle. These forms are scarce in some typical stools and mucus-flakes, in others they are tolerably

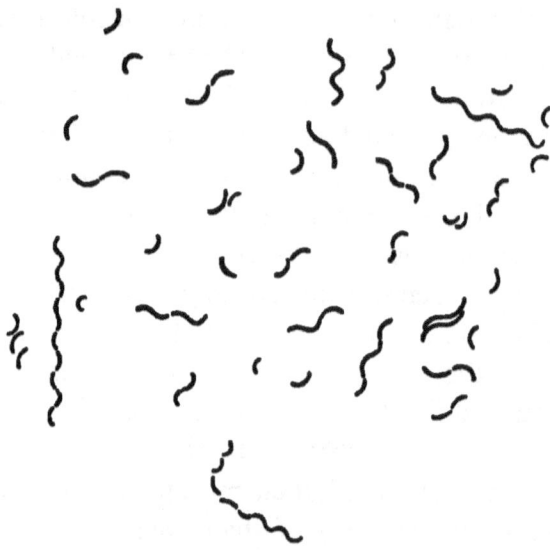

Fig. 6.—From a Preparation of Mucus-Flakes from the Lower Ileum, which had been allowed to undergo Putrefaction for three days. Magnifying power 700.

abundant. I have specimens of the stools of a patient ill with cholera a few hours only, in which the circular and semicircular forms were the only conspicuous forms; they were of two different sizes, some about half the size of others. Then I have specimens of the mucus flakes from the ileum of cases that died within the first day, in which these forms are very scarce, while in one case dead of typical acute cholera in $9\frac{1}{2}$ hours the number of small circular and semicircular forms is very conspicuous. Here also some were larger than others.

I have for a long time searched for an explanation of these forms, and I think I have found it. One cause of their abundant appearance seems to be this: the comma-bacilli, by transverse division, and by remaining joined end to end, occasionally form marked spirals, in which the elements are so much curved that the spiral possesses very close turns; when such a spiral is broken up into its constituent elements, we obtain semicircular and almost circular commas. That this is so I have ascertained in several instances, by comparing fresh specimens with dried and stained ones. While in the perfectly fresh specimens in these cases a good many spirals with closely-twisted turns were to be made out, after drying a thin film on the cover-glass, staining and mounting it, spirals were absent, being evidently broken up into the numerous semicircular forms then present. Another mode of the formation of these will be given below. Why there should be numerous close spirals in some cases and not in others I cannot say; but it seems to me that the solidity and resistance of the medium in which they grow has something to do with it, *i.e.* where the material in which they grow is and remains solid and owing to the resistance offered to the dividing comma-bacilli, these are limited in the area of their expansion; hence those which after continued division remain joined endways, become pushed more closely together, become more curved, and form spirals of closer turns. This, I am induced to think, is in some cases a true explanation; for not only do these forms abound in the depth of solid nutritive material, such as Agar-agar mixture, which remains solid, or in albumen-mixtures, but also in the solid particles taken from the ileum of typical cholera cases. I do not mean to say that other conditions do not determine the presence of these forms, for I shall show below that the age of a culture has also something to do with it.

All comma-bacilli when examined fresh show a rapid rotatory movement; this is obviously the more pronounced the more curved they are; it is very characteristic in the "dumb-bells," in which the two elements are so arranged that their curvature is directed in opposite ways, and in the chains of close or open spirals. The dumb-bells just mentioned are the result of the successive division of a comma-bacillus; they are generally S-shaped and are very characteristic and generally present along with the single commas. But some dumb-bells also occur in which the two elements are curved in the same direction, being more of the shape of ⁀⁀; the figure of a flying bird. In the fresh stools and fresh mucus-flakes chains longer than dumb-bells are not frequent, although, as mentioned above, they do occur isolated, and then chiefly in fresh specimens; in dried and stained specimens they are absent, evidently owing to the facility with which they are broken up into single elements and S-shaped forms during the act of preparation. Of these chains and spirals more will be said below.

The single elements are rounded at their ends; in some cases, however, a slight thinning at the ends can be made out with very high powers, such as a $\frac{1}{20}$ oil-immersion. In dried and stained specimens the comma-bacilli of the stools and intestinal contents appear uniformly stained, but on careful washing after staining, it can be shown that they consist of a delicate sheath with protoplasmic contents. One peculiarity the comma-bacilli possess is that the stain is easily taken out of them by alcohol, more easily than is the case with many other bacteria. On spreading out on a cover-glass a thin film of the mucus-flakes of the ileum, or of a rice-water stool, drying well over the open flame, then staining it

CHOLERAIC COMMA-BACILLI.

for five to ten minutes in a 2 per cent. watery solution of Spiller's purple, or Weigert's gentian-violet anilin-oil, then washing it in water, then just rinsing it once only with spirit, then with water, drying and then mounting in Canada balsam, many comma-bacilli of the typical lengths will be met with which show this distinction between a faintly-tinted sheath and the protoplasmic contents in the shape of two stained particles, one at each end, with a faintly stained interval, the lumps being rod-shaped and slightly curved; the curved state can be only made out with $\frac{1}{20}$ oil-immersion lens. I conclude from this that a single typical comma-bacillus is

FIG. 7.—PREPARATION OF CHOLERAIC COMMA-BACILLI STAINED WITH GENTIAN-VIOLET, AND AFTERWARDS WELL WASHED.

The differentiation between the sheath and protoplasm generally collected at the end of each comma-bacillus is well seen. Magnifying power about 1,400.

composed of two slightly rod-shaped elements held together in a common sheath. But there are numerous short comma-bacilli, which contain only a single rod-shaped protoplasmic element, situated at one end or occasionally also in the middle. The longer the comma-bacillus the longer the protoplasmic element. This enables us to say then, that the element is a protoplasmic granule, more or less rod-shaped, and according to its state of growth or elongation the comma-bacillus is longer or shorter; and further that when this element has by transverse division given origin to two protoplasmic elements, we have a comma-bacillus consisting of a common sheath, and in it, at each end, a protoplasmic particle; finally, when this sheath becomes divided transversely, the single comma-bacillus has divided into two comma-bacilli. The shortest comma-bacilli such as are present in all stools and mucus-flakes, and which particularly abound where rapid multiplication occurs, are composed of a sheath and a single protoplasmic element; the longer and typical examples contain two longer protoplasmic elements. The same structure can by careful staining and washing be ascertained also in the comma-bacilli of artificial cultures, not only of the choleraic ones, but also of other species of comma-bacilli, as will be mentioned below.

B.—*Comma-bacilli in Artificial Cultivations.*

The finest and most typical forms of comma-bacilli are obtained by placing, after the manner of Koch, a few mucus-flakes of the contents of the ileum of a fresh case of cholera on linen, and keeping this in a glass dish and under a bell-glass, on the inside of which a piece of moist blotting-paper has been fixed; in other words, in a moist chamber, at a temperature of about 20°-25° C. After twenty-four to thirty-six hours the comma-bacilli have enormously increased; numerous S-shaped and spirillar forms are met with. They differ in no respect from those described in the fresh intestinal contents. From an acute typical case of cholera, dead within twelve hours, in the ileum of which there were present large numbers of large and small mucus-flakes suspended in a little watery fluid, masses of mucus-flakes were taken and placed in a clean glass dish and covered up with a glass plate, and left standing for three days. Examination showed large numbers of various species of

CHOLERAIC COMMA-BACILLI.

bacteria, and also crowds of comma-bacilli. Comparing the comma-bacilli of such preparations with those made of the same flakes while fresh, a marked difference was noticed: in the former the comma-bacilli were obviously three or four times longer and thicker than in the latter, and besides there were present many long chains of comma-bacilli, wavy but not spiral. On growing the comma-bacilli in 10 per cent. gelatine and beef-broth, in plate-cultivations, after three or four days the comma-bacilli are very finely curved and all of about the same length, either single or S-shaped;

FIG. 8.—ARTIFICIAL CULTIVATION OF CHOLERAIC COMMA-BACILLI IN ALKALINE PEPTONE GELATINE.
Magnifying power 700.

in fact, the uniformly finest and most typical forms I have ever obtained were seen in plate-cultivations in (10 per cent.) gelatine and beef-broth. In 10 per cent. gelatine, beef-extract and peptone (1 per cent.), or gelatine (10 per cent.), beef-broth and peptone (1 per cent.), the comma-bacilli are also fine, but not so uniformly curved and typical as in the former medium. In alkaline or neutral beef broth during the first two or three days of growth at 35°-37° C., large numbers of very short comma-bacilli are met with amongst longer ones and S-shaped ones; but the curvature is well-pronounced only in few, most of them are only slightly curved. By this time there are also present a few long wavy and spiral forms; these latter increase in numbers as growth proceeds; the longest spirals, some extending over one-fourth or one-third of the field of the microscope (Zeiss's eye-piece 3, obj. D), I have found in alkaline broth-cultures that had been growing at 35° C. for a week. Some of them are spirally twisted, others more wavy, in some one portion is a spiral, while another portion is wavy, or apparently almost straight. In some well-washed stained specimens the individual commas constituting the wavy or spiral form can be recognised, in others there seems no

FIG. 9.—ARTIFICIAL CULTIVATION OF THE SAME COMMA-BACILLI AS IN PRECEDING FIGURE IN ALKALINE PEPTONE BROTH GELATINE.
The comma-bacilli are not so large and not so well curved.
Magnifying power 700.

differentiation to be made out. During the drying and staining of the specimens many long spirals break up into short spirals. In cultivations carried on in broth (alkaline) at 35°-37° C. a pellicle makes its appearance already after 36-48 hours, this thickens and becomes more connected as growth procceds; this pellicle is made up almost entirely of longer or shorter spirals.

In Agar-agar mixture, alkaline and neutral (Agar-agar 1 per cent., peptone 1 per cent., both dissolved in beef-broth), the comma-bacilli are not so well curved and uniform in size as in gelatine mixture. On growing the comma-bacilli in Agar-

agar mixture for from several days to several weeks at 35°-37° C., a good many spiral forms are met with; some have very close turns, others are more wavy, some contain as many as ten to twenty and more turns, others one or two; a good many S-shaped ones are also present.

In an alkaline mixture of egg-albumen and Agar-agar the comma-bacilli assume a very striking morphological character; they are all of a uniform appearance as regards length and thickness, but there is hardly any curvature to be noticed; that they are not quite straight is seen when they

FIG. 10.—ARTIFICIAL CULTIVATION OF THE SAME COMMA-BACILLI IN ALKALINE BEEF BROTH.
Magnifying power 700.

are in groups: looked at singly or situated at intervals many look only slightly or not at all curved, and it would not be easy to identify them as comma-bacilli. They are distinctly pointed at the ends, and longer than the normal ones. A preparation from a gelatine-culture compared with one from this medium shows apparently two totally different organisms; no one would recognise them as the same organism (see Fig. 13). Yet there can be no question about it, since on changing the medium by transferring them from nutritive gelatine into egg-albumen and Agar-agar,

E

or *vice versâ*, they change their morphological characters accordingly. I have examined numerous specimens from such a culture-tube of egg-albumen and Agar-agar, after several days' to several weeks' growth at $35°$-$37°$ C., and although there was copious growth present I did not see any spiral forms. The reason is obvious, namely, that since the individual comma-bacilli are almost straight, or at all events only very slightly bent, their chains are more like leptothrix: a spiral can of course only be produced by a chain of well-curved organisms; and just as a leptothrix is made up of a chain of straight bacilli, so a spiral is one of curved organisms. This holds good also for the S-shaped forms of the comma-bacilli; in the egg-albumen and Agar-agar mixture there are numerous dumb-bells, but they bear not much resemblance to the typical S-shaped forms of the comma-bacilli, since the elements in the former are more or less straight.

There can, then, be no doubt about the fact that in the different artificial media the comma-bacilli show distinctions as regards size and curvature. The most striking difference is that shown by comma-bacilli cultivated in an alkaline mixture of egg-albumen and Agar-agar.

Of very great importance, both from a morphological and from a physiological point of view, in the life-history of all bacterial species, is the question of spore-formation. Hitherto in those species only which are known as bacilli has real spore-formation been observed. In some species of bacillus the single rod or part of a chain or leptothrix-thread is capable of producing at the expense of its protoplasm bright glistening spores, spherical, or more generally oval in shape; and these spores when fully developed leave the sheath of the bacillus and represent the real seed, for they are capable of retaining vitality for an indefinite time, and when planted in suitable soil germinate again into rods, and these by elongation and transverse division give origin to two new bacilli, each of which continues to multiply by division. The formation of such spores, possible only under certain favourable conditions, such as free access of oxygen, suitable temperature, and moisture, constitutes the final step in the life-cycle of a bacillus, as it does in that of the higher fungi and higher plants.

Where owing to the nature of the bacilli or to unfavourable conditions (such as the absence of free oxygen in the

Fig. 11.—From a Cultivation of Choleraic Comma-bacilli in Liquefied Gelatine, after several weeks.

case of *Bacillus subtilis* and *anthracis*) spore-formation does not set in, the bacilli, having multiplied as long as the nutriment lasts, undergo finally a retrograde change, consisting in the breaking-up and disintegration and death of their protoplasm. This is probably due to the poisonous nature of chemical substances produced by the bacilli themselves. Such a culture ultimately becomes devoid of living protoplasm, and is incapable of starting new growths.

These conditions are well-known and have been studied

and ascertained by many competent observers, amongst whom particularly Cohn and Koch may be mentioned.

If we then find, as is the case with the various species of micrococci and *Bacterium termo*, that a culture of an organism, kept under the most favourable conditions for the formation of spores, loses after some time the power of starting a new crop in a suitable medium, we are justified in saying that in such a culture no living or life-giving particle is present, no spore has been formed. Looked at from this point of view, I am in agreement with Koch, who from experimental observations (described on a former page) denies the formation of spores in the comma-bacilli and spirilla. I have had during the last four years a large number of culture-tubes of comma-bacilli in gelatine and Agar-agar mixture, which after several months proved barren of all life, although they were once good and active cultures. I have had cultures made in Bombay during September 1884, which for some months contained a copious crop of living choleraic comma-bacilli. After my return to England in January 1885, they were tested and were found capable of starting active and good fresh cultures of comma-bacilli. And so they were found till May 1885, *i.e.* after nine months. But after this time all life in them became extinct. Subsequent experiments carried on on a large scale have confirmed this; many tubes tested in this respect were, before the end of twelve months or earlier, proved barren of all life. As regards gelatine-tubes, such a condition, *i.e.* death of the growth, sets in in many instances after four to five months, in others after six to eight months. (Compare also E. Weibel's observations of other vibrios.)

Now from a morphological point of view it can be easily ascertained that in almost all culture-tubes, say in gelatine tubes after two to three weeks, in Agar-agar tubes kept at

35°-37° C. earlier, many comma-bacilli and spirilla are met with in which the protoplasm has become granular, and they can be traced into forms in which the granules have become free, leaving the pale sheath behind. A careful examination leaves no doubt that these granules which in old cultures are met with in masses, are the *débris* of dead organisms. Hueppe describes the presence in the single comma-bacilli,

FIG. 12.—PREPARATION OF A PURE CULTIVATION OF CHOLERAIC COMMA-BACILLI IN AGAR-AGAR MEAT-EXTRACT PPETONE, SEVERAL MONTHS OLD.

1. Semicircular forms.
2. Circular forms.
3. Spirilla.

Magnifying power about 1,400.

and in the spiral forms, and in a free state, of granules which he maintains to be spores (Arthrospores), having, he states, observed them in a few instances to germinate again into comma-bacilli. Bearing in mind that, as we have already shown (p. 45), the comma-bacilli in all forms show at the outset a differentiation between protoplasmic contents and sheath, and that the size, or rather length, of these protoplasmic elements depends on the phase of growth, part of the above assertion of Hueppe—that those elementary masses of protoplasm are capable of elongating into well-formed comma-bacilli—is perfectly intelligible, without any need for ascribing to them the character of spores. The above observation of the final death of the cultures seems to me to prove that they are not spores. The morphological test alone is therefore unsatisfactory; that is to say, the presence of elementary or young masses of protoplasm in the single comma-bacilli, or in those forming a spirillum or in a free state, does not seem to me conclusive proof that

Fig. 13.—Preparation of a Cultivation of Choleraic Comma-bacilli in Egg-Albumen and Agar-Agar, ten days old.
The comma-bacilli show only a slight curvature, many of them are almost straight; all of them are distinctly pointed at the ends. Magnifying power about 1,400.

they are spores in the same sense as are the well-known and well-characterised spores of, say, *Bacillus subtilis* or *Bacillus anthracis*, since we recognised that condition as both intimately bound up with the normal structure of a comma-bacillus and as the result of disintegration.

With regard to the alleged spore-formation in the cholera-bacilli, Koch says (*loc. cit.*):—"As the question of the presence of a resting-form of cholera-bacilli is down on our programme, I will say a few words with regard to it. On account of the importance of this question, I have done my utmost to find something which could be looked on as a resting-stage of the cholera-bacilli, analogous to the spore-formation of other bacilli, but I have, just as in my former investigations on this point, obtained only negative results. All the statements as yet made by other observers on the resting-forms and spore-formation depend evidently on errors." (Compare also Koch's experiments mentioned on pp. 16 and 17.)

The complex process of spore-formation of the choleraic spirilla described by Ferran, namely by the formation of a peronospora, has not been confirmed; and it is difficult to come to any other conclusion as regards them than that Ferran's cultivations were impure, and that accompanying

FIG. 14.—FROM AN ARTIFICIAL CULTIVATION OF CHOLERAIC MUCUS-FLAKES ON DAMP LINEN.
Comma-bacilli of the ordinary type. Amongst them are some much thicker containing a vacuole.
Magnifying power about 700.

the comma-bacilli there were probably present other forms as an accidental contamination, or that he mistook degenerative forms for indications of spore-formation.[1]

A curious degenerative change, besides the granular disintegration already mentioned, is observed in some Agar-agar tubes, consisting in the swelling up of one end of the comma-bacilli, whereby the ordinary comma-bacilli become changed into pear-shaped or club-shaped, straight or more or less curved and twisted, corpuscles. But these ultimately

[1] Carillon, Ceci and others describe spherical enlargements and vacuolation as spore-formation in the comma-bacilli.

undergo granular disintegration, so that their pale sheaths alone are left.

An interesting modification is noticeable in the comma-bacilli in the stools and mucus-flakes taken from the ileum and grown on linen, by placing mucus-flakes on a piece of linen and keeping this in a capsule under a bell-glass, inside of which is a piece of wet blotting-paper, at a temperature of 20°-25° C. On examining microscopic preparations after twenty-four, thirty-six, or forty-eight hours, the comma-bacilli are found to have increased very greatly, and amongst the

Fig. 15.—Preparation of a Cultivation of Choleraic Comma-bacilli on Damp Linen after Thirty-six Hours.
1. Typical comma-bacilli.
2. Vacuolated comma-bacilli.
3. Circular and oval forms produced during the process of vacuolation.
Magnifying power about 1,400.

typical forms—single commas, S-shaped and spiral forms—numerous examples are found which are conspicuously thicker, and include smaller or larger vacuoles, singly in the centre or in a row of two or more. According to the size of this vacuole the comma-bacilli are no longer convex concave, but plane-convex or bi-convex. If the vacuole is large they appear more or less circular. There can be no question about this change, since almost in every field we find every intermediate form between a typical comma-bacillus of the normal thickness and a circular or oval form.

CHOLERAIC COMMA-BACILLI.

In a stool from an acute case of cholera I have seen these forms in enormous numbers; in some fields of the microscope the transitional forms between complete circles and typical commas being very numerous. In neutral Agar-agar mixture kept at the temperature of the room (between 18° and 20° C.) after three, four, and more weeks, I have seen this change occurring in many of the commas; in alkaline Agar-agar mixture I have also noticed it, but more rarely. The best specimens I possess were made from cultures in neutral Agar-agar about four to six weeks old. Preparations made of such cultures show in some fields of the microscope the typical comma-bacilli only in a few

FIG. 16.—FROM A RECENT ARTIFICIAL CULTIVATION OF CHOLERAIC COMMA-BACILLI IN ALKALINE AGAR-AGAR JELLY.
Magnifying power about 700.

examples, most of them showing in some stage or other the above vacuolation and transformation into circular or oval organisms. Some of these, although already completely rounded, nevertheless do not yet show in their interior a complete cavity, but two or three small cavities separated by remnants of protoplasm.

As the comma-bacilli from which these rounded or oval forms take their origin are of different lengths, so do also the sizes of these rounded or oval forms differ, the small commas giving origin to small, the large commas to large, rounded or oval forms. The question is, whether these rounded forms are discoid or spheroidal. I am inclined to think that some of them, at any rate, are discoid, because in the fresh state, and suspended in fluid, I have noticed under the microscope that when they float and roll they

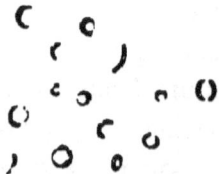

Fig. 17.—From an Artificial Cultivation of Choleraic Comma-bacilli on Neutral Agar-agar Jelly at Ordinary Temperature, after a few weeks. Some of the comma-bacilli have become converted into circles, which by division give origin to two semicircular comma-bacilli.
Magnifying power about 700.

present alternately a broad and narrow surface; but there is evidence that some are distinctly spheroidal, since when rolling they always show a circular outline. The next and important question is: What is the meaning of this change?

Fig. 18.—From a Similar Preparation.

Fig. 19.—From a Similar Preparation.

Koch, in a somewhat impatient way, says that these forms described by me (which, as will be shown presently, are the first stages of a longitudinal division) are a supposed discovery "resting on an erroneous interpretation of the involutional (*i.e.* degenerative) forms of the cholera-bacilli."

I do not think Koch has taken the trouble to examine more carefully what I did say, for I distinctly stated that some of these forms when examined fresh show as active a movement as the comma-bacilli themselves. Surely this proves that they are living. And by careful examination it can also be shown that in many of these forms a single or double thinning leading to deficiency of the wall of the round organism is present, by which the organism divides into two more or less semicircular comma-bacilli.

In mucus-flakes and fresh stools from some acute cases such forms—namely completely circular organisms, and circular organisms with one or two gaps in the wall at opposite poles—are very numerous, and from the above there can be little doubt that these owe their origin to the same process of vacuolation and subsequent division. I have got specimens of linen-cultures, of fresh stools, and of Agar-agar cultures where the whole chain of changes is so marked and so well illustrated that on a careful examination one cannot help arriving at the conclusion just stated. Some of these circular or oval forms show even three gaps or fractures, so that one circle gives origin to three comma-bacilli. And, finally, there are some in which the whole protoplasm thins, and ultimately breaks away and disappears, except a short rod or granule at one spot of the circumference, while at the same time a discoloured or circular pale sheath remains. I have before me specimens made from fresh and active linen-cultures, where almost all and every comma-bacillus is thus changed; to speak of such specimens as indicating degenerative changes seems to me unwarranted. These forms are well shown in Fig. 15.

On the whole, then, I think I am justified in maintaining that these circular and oval forms represent the initial stages of a mode of division differing from the ordinary mode of transverse division. This ordinary or typical mode, as observed in various bacilli, consists in the elongation of a single bacillus, and the subsequent transverse fission of this into two new rod-shaped elements. That this is also the ordinary mode of propagation of the comma-bacilli is proved (*a*) by the different lengths in which comma-bacilli occur, (*b*) by the S-shaped forms, (*c*) by the chains, wavy and more or less spiral, representing in reality a series of comma-bacilli placed end to end. But in our instances we have to deal with a mode of propagation essentially different from the above; for here a single comma-bacillus does not elongate, but by vacuolation becomes thicker, and changes finally into a more or less rounded or oval form, giving then origin to two new comma-bacilli. This mode may be described as the *atypical* mode of division, or *division after vacuolation*.

CHAPTER IV.

CHARACTERS OF THE COMMA-BACILLI IN ARTIFICIAL CULTIVATIONS.

THE choleraic comma-bacilli possess in artificial media certain well-established characters, by which they can be readily recognised. Koch has clearly pointed out this fact, and has minutely described the appearances. An idea seems to have got abroad that while in India I denied this simple truth. I said then that the choleraic comma-bacilli do not differ in this matter of artificial cultivation from other septic bacteria; and I still say so now, after admitting and confirming the correctness of Koch's observations as to the behaviour of the choleraic comma-bacilli in nutritive gelatine. No one who has carefully followed recent research, and has sufficient practical experience in the artificial cultivation of bacteria, can for a moment doubt that the different bacteria, be they septic or pathogenic, have morphological and cultural characters of their own, which in some instances are more, in others less, pronounced; but a good many species are known in which these differences in mode of growth are sufficiently striking to be of diagnostic value even to the unaided eye. Take, for instance, micrococci derived from the air. There are a good many species of micrococci present in the air of ordinary city laboratories, which can be made visible by simply exposing a number of sterilised glass dishes, in which previously a thin layer of nutritive gelatine or Agar-agar mixture liquefied by warmth has been poured out and allowed to set over a layer of cold water. Each glass dish is kept covered up with another glass dish, and the whole kept in a moist chamber, or in other words on a glass plate covered up by a bell-glass to which a piece of wet blotting-paper has been fixed. This for all practical purposes represents a good plate-cultivation after Koch's method. On exposing for from several seconds to several minutes the solidified layer of the nutritive medium, by lifting off the upper covering glass dish, then closing it again and placing the bell-glass in position and keeping the whole at a temperature of 19°-22° C. in the case of gelatine, or 35°-36° C. in the

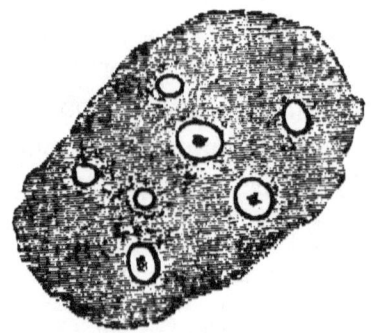

Fig. 20.—Plate-Cultivation in Gelatine 48 hours old, showing Young Colonies of Choleraic Comma-bacilli.

case of Agar-agar mixture, for two, three, or more days, various spots or colonies will show themselves on the surface of the plate-cultivation, some small, others large, some round, others irregular, some spherical, others disc-shaped, some whitish or greyish, others yellow, orange, or of other colours, some not liquefying the gelatine, others liquefying it ; of the latter some liquefy it with a smooth circular outline, others with an irregular serrated line; in some the liquefied gelatine is clear throughout the circumference and extent of the colony except for a central opaque speck; in others the liquefied area is uniformly turbid, and so on. By careful examination and re-inoculation of nutritive gelatine or Agar-agar on plates and in tubes the different species characterised by the different appearances just named can be isolated and studied.

The same can be observed in the case of bacilli and bacteria, *i.e.* a different mode of growth as regards rapidity of increase or size of the colonies after a certain time, and as regards colour, aspect, outline, and microscopic characters of the organism constituting the colonies. To say, therefore, that such or such an organism in plate-cultivation and in tubes presents such and such peculiar characters, means nothing more than that such and such an organism is of a definite species, and, as we have said, the greater majority of the bacterial species are possessed of such special characters. I never said that the choleraic comma-bacilli cannot in cultivation be distinguished from other bacteria. If any one thinks I did, I can only answer that he has misunderstood my meaning, and at the same time has failed to apprehend the simple fact demonstrated by Koch himself, that almost all the different species of bacteria show under cultivation different cultural characters, by which they are more or less easily distinguishable one from another. I say the possession of cultural characters is not peculiar to comma-bacilli. This is something quite different from saying that the cultural characters of comma-bacilli are the same as those of septic bacteria.

What are, then, the characters shown by the choleraic comma-bacilli in artificial cultivations?

On a former page it has been mentioned that very successful cultivations of comma-bacilli can be made by placing mucus-flakes of the rice-water stools or of the cholera intestine on linen, and keeping this damp in a covered glass dish at 35°-37° C. After twenty-four hours crowds of comma-bacilli are available, from which plate-cultivations can be then easily made.

But this is only the case if the mucus-flakes initially contain the comma-bacilli in considerable numbers. Another method is this: with a particle of a mucus-flake or with a minute droplet of the rice-water stool, a test-tube containing sterile salt solution or sterile broth is inoculated, well shaken, and from this a tube containing sterile solid nutrient gelatine is inoculated, liquefied in warm water, and from it one or more plate cultivations are established; or if the comma-bacilli are tolerably numerous in the stools a tube containing sterile broth is inoculated with a trace of those materials and then incubated at 35°-37° C. Owing to the rapid power of growth of the comma-bacilli, after twenty-four hours the fluid contains them very abundantly, and now a tube containing sterile salt solution or broth is inoculated, well shaken, and from it nutrient gelatine is inoculated with a trace of a droplet (with the end of a platinum wire or capillary pipette); this is liquefied and used for plate-cultivations. As will be stated further below, when other bacteria are numerous but the comma-bacilli scarce in the original materials (stool or mucus-flakes) a number of plate-cultivations (after the usual method of dilution) will have to be made in order to obtain a few colonies of the comma-bacilli.

The appearances presented by these comma-bacilli when grown in gelatine plates, in gelatine tubes, in Agar-agar mixture, and on potato, in broth, milk, &c., are sufficiently characteristic for all practical purposes of diagnosis.

(*a*) *In broth.*—Faintly alkaline beef-broth, or broth to which 1 per cent. of peptone is added and rendered faintly alkaline, or meat-extract peptone (1 large tin of Brand's meat-extract, 10 grms. of peptone in 1000 ccm. of distilled water made faintly alkaline) are very good fluid media, and in them the comma-bacilli multiply with great rapidity. A flask or test-tube of these materials inoculated with the comma-bacilli and kept in the incubator at $35°$-$37°$ C. is already slightly turbid in twenty-four hours, every drop contains multitudes of single commas and S-shaped forms. After three to four days the turbidity has greatly increased, and a slight powdery precipitate is noticed. On the surface of the broth a thin loose pellicle is noticed already after 2–3 days, this becomes more complete as growth proceeds; on shaking the tube the pellicle easily breaks up into flakes or thin scales, many of them remaining on the surface. The fluid remains thin, but has distinctly changed its reaction, having become acid; there is no offensive smell, but rather a slight aromatic flavour. The fluid cannot yet be considered exhausted, since even after a week the turbidity remains unaltered, while the precipitate increases. After about a fortnight the fluid begins to clear, as it were in layers, beginning at the surface and gradually extending to the depth. The precipitate meanwhile increases, and the fluid becomes after the lapse of weeks almost clear (in the upper part) to the unaided eye, but on examining a droplet under the microscope moving comma-bacilli, singly, but chiefly S-shaped and spiral forms, can be easily detected. The precipitate, from its first appearance, is made up of living comma-bacilli and of the granular *débris* of dead ones; in these the outline and sheath can in many instances still be recognised, either with or without granules; besides these there are smaller and larger clumps and masses of granular *débris*.

The comma-bacilli grow equally well in neutral broth. I have seen them grow fairly well even in faintly but distinctly acid broth, but the amount of turbidity and of precipitate was far less than in alkaline and neutral broth.

(*b*) *In milk* the comma-bacilli grow well, but not so luxuriantly as in broth; the appearance of the milk (as has

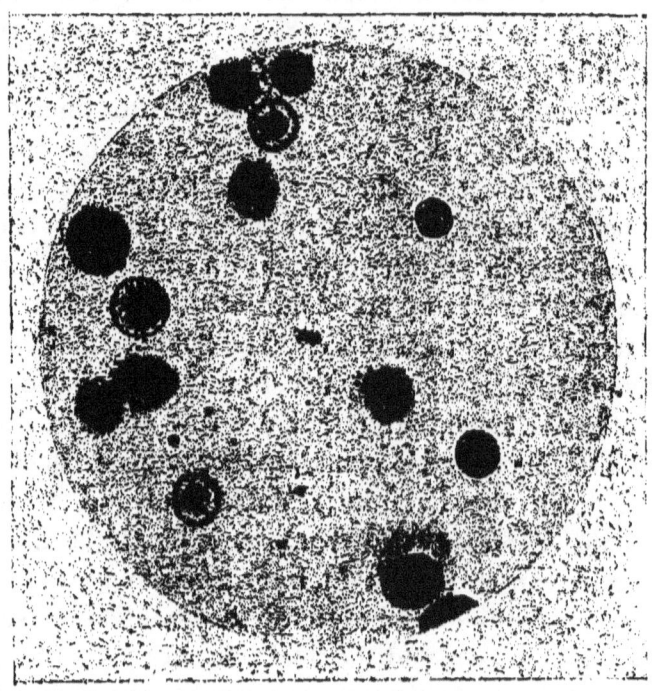

FIG. 21.—PLATE CULTIVATION WITH COLONIES OF CHOLERAIC COMMA-BACILLI, SEVENTY-TWO HOURS OLD. MAGNIFIED FIFTY TIMES. FROM A PHOTOGRAPH AFTER PLAGGE.

been pointed out by Koch) remains practically unaltered, and the casein is not precipitated. Mr. Warrington (*Journal of the Chemical Society*, 1888) has noticed curdling of milk to take place by the comma-bacilli when growing at 30°C.

(*c*) *In Agar-agar mixture* (Brand's meat-extract 1 tin, dissolved in 100 ccm. of distilled water, to which is added 10 grms. of Grubler's peptone, made faintly alkaline, boiled and filtered; add to this a filtered boiled solution of 10 grms. of Agar-agar in 900 ccm. of water). Faintly alkaline beef broth or beef infusion, to which are added 1 per cent. Agar-agar and 1 per cent. peptone, is also very good.

The comma-bacilli, sown at a point of the surface, grow at 35–37° C. in the course of a few days into a translucent film, with rounded or slightly indented outline; this film gradually enlarges in extent and becomes more or less terraced, *i.e.* arranged in superimposed layers differing in extent; radiating lines probably due to shrinking of the Agar-agar mixture are now to be noticed; the central portion

Fig. 22.—From a Cultivation of Choleraic Comma-bacilli in Gelatine in a Glass Dish four days after Inoculation in Spots. Natural Size.

is thickest and therefore more opaque than the periphery; the whole growth after a few weeks to a few months shows in transmitted light a yellowish-brown tint with denser brownish spots, in reflected light it looks whitish grey. In streak cultivation a film starts from the line of inoculation which rapidly (in two to three days) spreads out in breadth.

Sown into the depth of the medium, into a channel made with the pointed end of a capillary glass tube or the platinum wire, and kept at a temperature of 35–37° C., the growth is always noticed after two or three days as a whitish line; this gradually increases in thickness, and the growth seen under a lens appears more or less granular.

(*d*) *In vegetable albumen and Agar-agar mixture* (Grubler's vegetable albumen, such as the white substance of Brazil nut, extracted with dilute alkali, filtered, precipitated, and then dissolved in dilute alkali; to the solution is added Agar-agar in the proportion of 1 p. c.) forms a translucent solid nutritive medium.' The comma-bacilli grow in this medium fairly well, but not so well as in Agar-agar and meat-extract peptone. They form a thin transparent greyish film on the

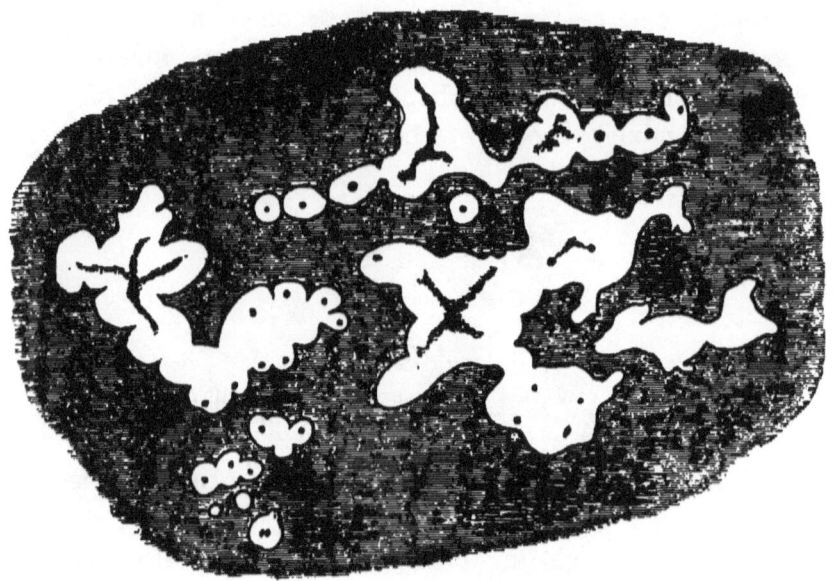

FIG. 29.—PLATE-CULTIVATION OF CHOLERAIC COMMA-BACILLI IN GELATINE AFTER FOUR DAYS AT 19° C. NATURAL SIZE.
Many of the colonies have become confluent.

surface with an irregular outline; when sown into the depth they form a greyish line indicating the channel of inoculation. But in neither case does the growth even after several weeks reach considerable dimensions, and in this respect the medium is very inferior to the Agar-agar and meat-broth peptone.

(*e*)*Egg-albumen and Agar-agar* is much preferable and is

nearly as good as Agar-agar meat-broth peptone. The white of an egg, about 25 to 30 ccm., is dissolved in 220 ccm. of distilled water to which 30 ccm. of liquor potassæ is previously added. Boil till all is quite dissolved; and after this add 4 grms. of acid phosphate of potassium, whereby the alkalinity is reduced but not quite neutralised; then add 1 p. c. of Agar-agar; dissolve by boiling, filter, and decant into test-tubes, which for sterilising are treated in the usual manner. This egg albumen and Agar-agar mixture is of good solid consistency even at 50° C., is beautifully transparent, and is a very good solid nutritive medium for the comma-bacilli, besides being very much cheaper (one egg being greatly cheaper and easier to obtain than a pint of beef broth or a tin of Brand's meat-extract) than the Agar-agar meat-extract peptone. In this medium the comma-bacilli grow in the same manner as in the Agar-agar meat-extract peptone mixture.

(*f*) *On linen.*—As Koch has pointed out, if the mucus-flakes from the ileum of an acute case are placed on linen, kept damp in a closed glass chamber at any temperature between 20° and 36° C., a very excellent cultivation of comma-bacilli is obtained. On examining these mucus-flakes after twenty-four to thirty-six hours or a little later, crowds of comma-bacilli are seen. In a fresh preparation many of these are wavy or spiral chains or S-shaped forms, but on making a permanent specimen by drying and staining most of them are found broken up into single comma-bacilli. T. R. Lewis has pointed out this difference (Appendix to *Report on Asiatic Cholera*). After forty-eight hours, owing to overgrowth of other bacteria present, the comma-bacilli cannot be easily obtained pure; but if originally the mucus-flakes contain few other bacteria besides comma-bacilli, a linen cultivation such as that described is a ready and excellent method of obtaining the comma-bacilli in almost a pure cultivation, or at any rate sufficiently pure to prepare from it successfully pure cultivations either in plates or in test-tubes. But if at the outset other bacteria are present in great numbers, one's success in obtaining anything like a satisfactory cultivation of comma-bacilli in test-tubes is doubtful. (See above.)

(*g*) *In gelatine* the comma-bacilli show the best-marked characters. Proceed after Koch's method by inoculating a test-tube containing nutritive gelatine (beef-broth or beef-infusion, gelatine 10 p. c., peptone 1 p. c., common salt 1 p. c.,) with the platinum wire or the pointed end of a capillary glass pipette charged with a tiny particle of a mucus-flake from the cholera-stool or from the contents of the ileum of an acute case of cholera (or—what for the purpose of class-demonstration is much easier—with a trace of a culture containing comma-bacilli pure or impure); then liquefy the gelatine and shake it gently but sufficiently so as to distribute well the introduced germs, and pour it out on a sterilised glass or plate, or better still, into a flat-bottomed sterilised glass dish sufficiently large to allow the gelatine to spread out into a thin layer; cover this immediately with a glass plate or glass dish, and let the gelatine rapidly set over ice or cold water, or by placing the glass dish on stone or metal in a cool place; place the dish in a moist chamber under a bell-glass as mentioned in a former page, and keep it at a temperature between 18° and 22° C. Within this range of temperature the comma-bacilli develop sufficiently well, while the gelatine remains solid.

After two, three, or four days, according to the tempera-

(If the number of comma-bacilli and other bacteria should be great, it is best first to distribute a particle of the mucus-flake in sterile salt solution or broth, and from this then to inoculate gelatine for making plate-cultivations, as described on a former page.)

ARTIFICIAL COMMA-BACILLI.

ture, the first indications of the colonies of the comma-bacilli make themselves visible as small round depressions somewhat greyish in colour. In another day each colony is a circular or oval, translucent or pearly-looking speck from the size of a millet seed to that of a small pea; the smallest when looked at obliquely still show the central pit, surrounded by a

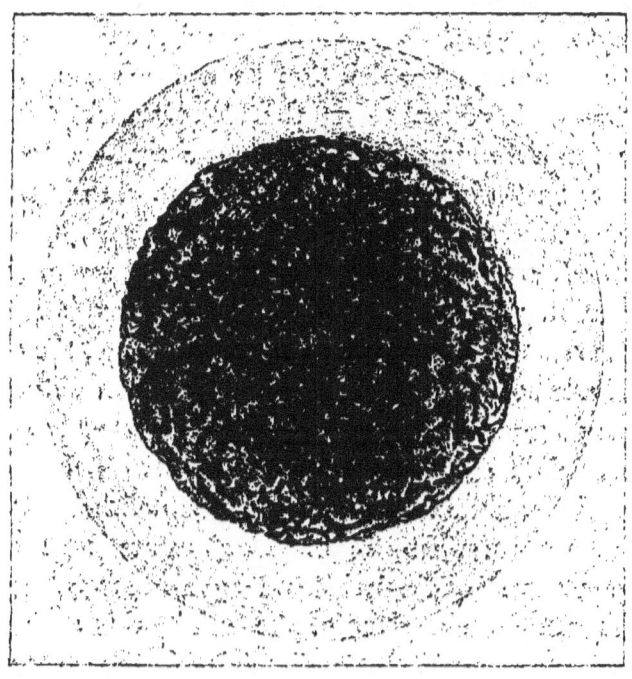

FIG. 24.—A COLONY OF CHOLERAIC COMMA-BACILLI IN GELATINE, SEVENTY-TWO HOURS OLD. MAGNIFIED 170 TIMES. FROM A PHOTOGRAPH AFTER KOCH.

greyish areola of liquefied gelatine, the larger ones showing no central pit but a white granule in the centre, while the areola of liquefied gelatine is greyish and translucent. Under a magnifying glass this greyish liquefying colony appears more or less uniformly and finely granular. In another day, while the colony enlarges in diameter—considerably larger than is stated by Koch, its outline is still smooth, circular or oval, and just at the margin marked by a whitish line that looks, on magnification, granular. The central opaque spot enlarges, and is granular, and gradually shades off, as it were, into the clear areola. In some plate-cultivations the outline of such colonies is not smooth, but more or less serrated. (Mr. Watson Cheyne incorrectly describes the irregular outline of the colonies as of constant character.) It is not correct to say that the colonies have always a serrated or irregular outline, nor that they have always a smooth outline. Nor is it correct to say that the colonies at their first appearance have already a central white spot—a precipitate of comma-bacilli in the centre—for I have seen in the same plates colonies which were of the size of a pea, but did not show the central spot, while others that were smaller possessed it. If the number of colonies that make their appearance in a given plate-cultivation is large, the contiguous colonies soon become confluent, and then we obtain an appearance something like that in Fig. 23, when the outlines of the original circular or oval colonies with their central spot and clear areola are still easily distinguishable. If the colonies were originally of irregular outline, by their confluence a correspondingly altered appearance is produced. But the general character of the colonies is their central depression and granulation and their peripheral more or less translucent areola of liquefied gelatine. When the growth has proceeded far enough and the original number of colonies is sufficiently great, the whole plate cultivation will be found liquefied in four or five days. There is in such liquefied plates a sediment of greyish powder, and here and there an attempt at something like a loose filmy pellicle; this latter does not, however, extend over the whole surface of the plate, but is only present here and there in the form of greyish scales. On examining under the microscope a droplet from any part of an early or advanced colony, there are always seen multitudes of actively moving comma-bacilli single or S-shaped and in spirals. A trace taken from the central granular spot or precipitate shows multitudes of S-shaped, wavy, or spiral forms more or less intimately matted together in larger or smaller masses. The spirals on drying and staining easily break up into single commas and S-shaped forms.

Owing to the comparative slowness with which the colonies of comma-bacilli in gelatine plate-cultivations make their appearance, it is clear that if other more rapidly growing bacteria, micrococci, and bacilli, which by their growth are capable of liquefying the gelatine, have been introduced in large numbers into the cultivation, the colonies of the comma-bacilli will be difficult to demonstrate. This is especially the case when one wishes to demonstrate by gelatine plate-cultivation the presence of comma-bacilli in a choleraic stool or in the contents of the ileum of an early cholera case in which the comma-bacilli are originally only sparingly present and are accompanied by multitudes of other bacteria; and we have already mentioned that this is not at all a rare thing, but on the contrary is more common than the reverse. In such cases the demonstration of the colonies of comma-bacilli in gelatine plates is not very easy of achievement. It is therefore necessary in such cases to dilute considerably with sterilised neutral fluid—salt-solution or broth—the particle of matter taken from the stool, and from this dilution to inoculate with a droplet the gelatine tube. It is obvious that owing to the relatively small number of comma-bacilli originally present it is necessary for success to start at the same time a series of plate-cultivations. I have seen cases of cholera where out of a dozen of plate-cultivations made by this method, only one showed evidence of a colony of choleraic comma-bacilli. The demonstration by plate-cultivation therefore of the presence of choleraic comma-bacilli in a given sample of a cholera-stool, or of the contents of the ileum, is in many cases not such a simple matter as is represented by Koch

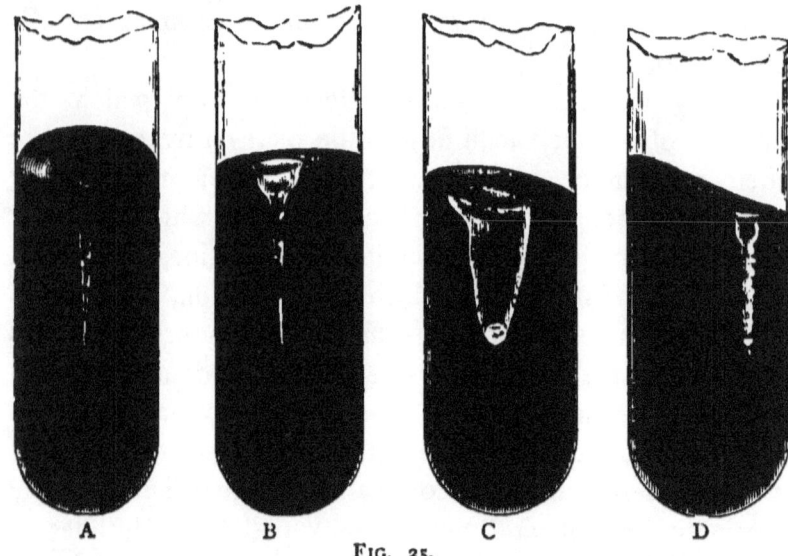

FIG. 25.

(A) Cultivation, in solid gelatine peptone broth, of straight mobile bacilli from the fluid of the mouth after four days' growth, showing a funnel-shaped depression, the lower part of the funnel filled with liquefied gelatine containing the growth of the bacilli. Semi-profile view.

B) Same tube viewed in profile.

(C and D) Cultivations in alkaline gelatine of choleraic comma-bacilli after five days' growth. In both tubes the inoculation had been made within a few seconds from the same stock. The surface shows the well-known depression; the channel in which the inoculation was made contains the growth of the comma-bacilli, the gelatine is here liquefied. At the bottom of the channel is a whitish precipitate of masses of comma-bacilli.

and others, for its success depends in a great measure on the relative number of comma-bacilli originally present. In some cases of undoubted cholera the result of such an examination is negative, while in others it is achieved only by making numerous plate-cultivations at the same time. It is however true that in some cases, namely where the

ARTIFICIAL COMMA-BACILLI.

comma-bacilli are originally present in very large numbers in the mucus-flakes of the rice-water stools, a couple of plate-cultivations reveal the presence of numerous colonies of the comma-bacilli.

In nutritive gelatine contained in tubes the comma-bacilli grow in a typical manner. On inoculating by first dipping

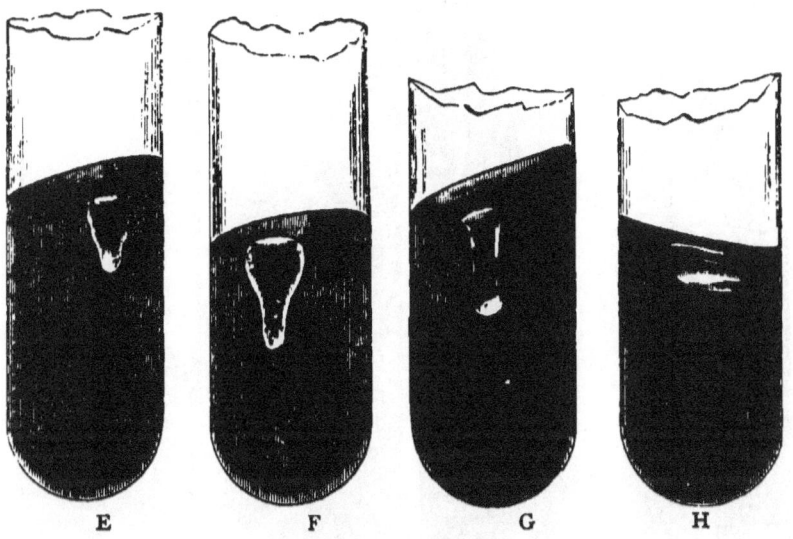

FIG. 25 *continued.*

(E) Cultivation of comma-bacilli from the fluid of the mouth (healthy).
F) Cultivation of choleraic comma-bacilli. In both E and F the medium is the same (alkaline gelatine peptone broth), and the inoculations were made within a few minutes.
(G) Cultivation of mouth comma-bacilli in gelatine peptone broth.
(H) Cultivation of choleraic comma-bacilli in alkaline gelatine peptone broth; the inoculation was made on the surface.

the platinum wire or the pointed end of a capillary glass pipette into a pure culture of comma-bacilli, and then pushing it into the (solid) nutritive gelatine, and exposing the tubes to a temperature of about 20° C., pure cultivations of the comma-bacilli are obtained in a few days. After the lapse of a couple of days the channel of inoculation becomes marked as a greyish line, this as growth proceeds broadens, particularly near the surface; the surface itself then shows a drawing inwards or depression of the gelatine, and as development proceeds this pit is converted into a

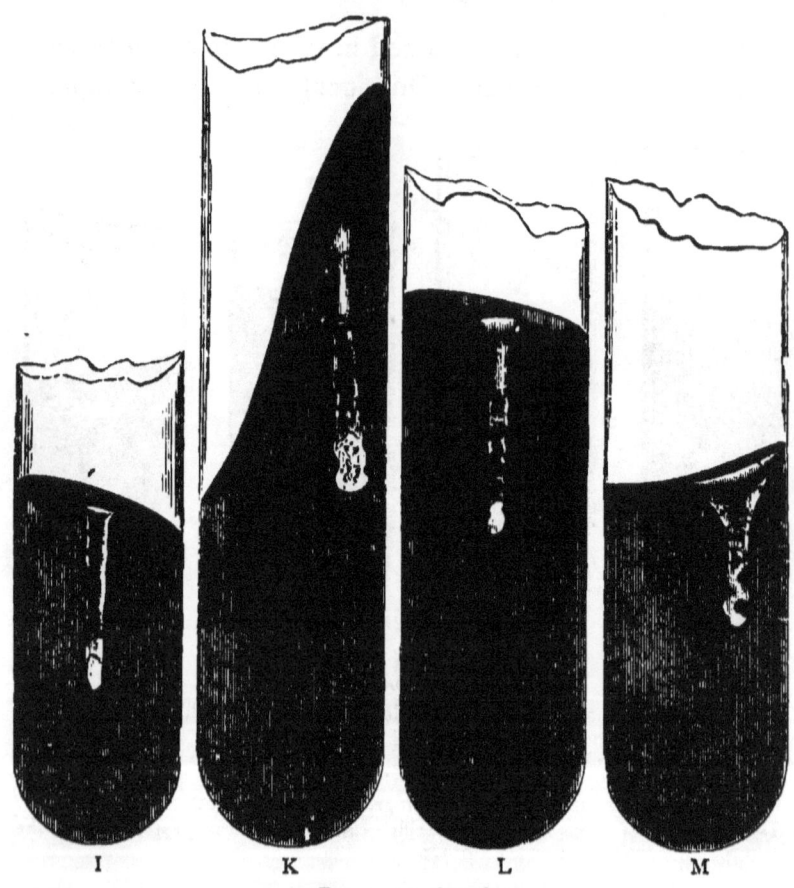

Fig. 25 *continued*.

(I) Cultivation of choleraic comma-bacilli in alkaline gelatine peptone broth.
(K L and M) Cultivation of choleraic comma-bacilli in alkaline gelatine peptone. In K and L an air-bubble occludes the upper end of the channel of inoculation.

funnel the mouth of which is occupied by an air-bubble. The rest of the channel of inoculation is a greyish canal filled with translucent liquefied gelatine, and at the bottom of the canal is a whitish precipitate. During the subsequent days the canal of liquefied gelatine broadens, the precipitate increases, and at the end of ten to fourteen days the

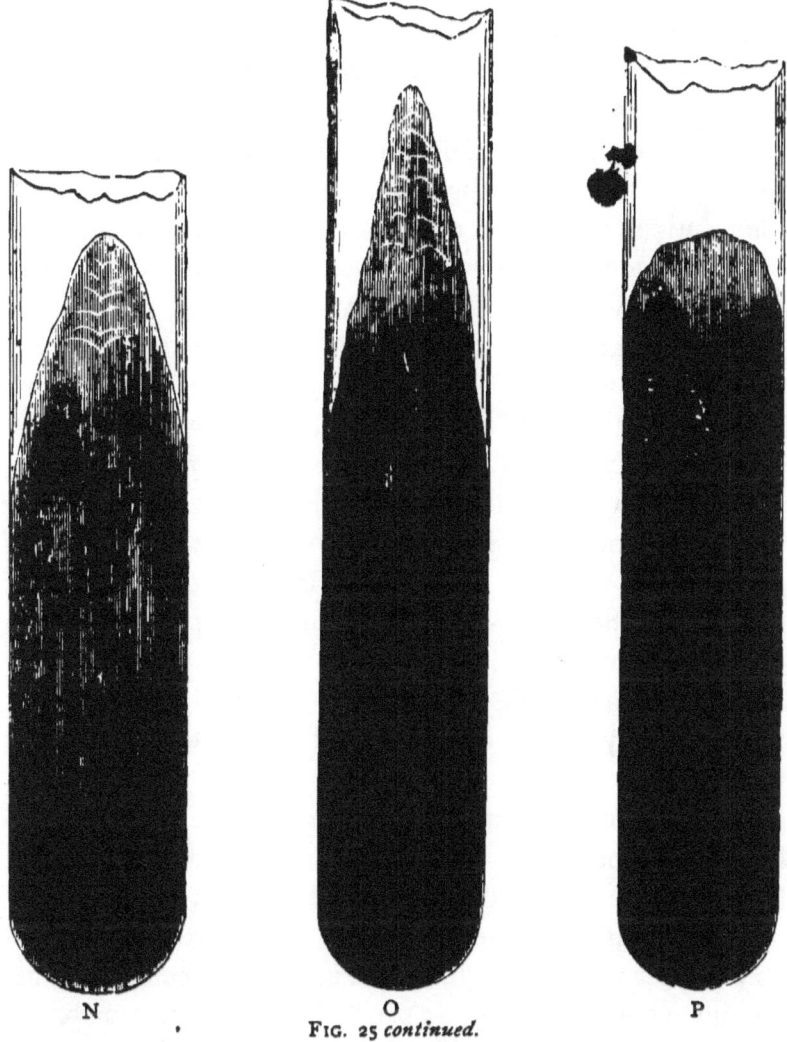

FIG. 25 *continued.*

(N O and P) Cultivation of choleraic comma-bacill in alkaline Agar-agar peptone and meat-extract.

growth has the form of a wide cone or a cylinder filled wit liquefied gelatine ; not only is there prseent a voluminous whitish precipitate at the deep end, but also at the sides of the liquefied gelatine there are present small granules easily recognised under a lens, while the bulk of the liquefied gelatine is translucent and almost clear. On the surface the funnel-shaped depression occluded by an air-bubble is still noticeable, but gradually diminishes and disappears as the growth extends more and more laterally. The growth at the end of three to five weeks has almost entirely liquefied the gelatine through the whole depth of the original channel of inoculation, and now the liquefaction proceeds deeper and gradually invades the deepest parts of the gelatine, the lower boundary of liquefaction being always marked by a voluminous whitish precipitate. While the bulk of the liquefied portion is tolerably clear there is present in most tubes on the surface a kind of loose whitish film; indeed I have seen very few tubes in which the surface remained free from it. Under the microscope the film is composed of granular *débris*, and moving spirilla more or less matted together. The above-mentioned funnel-shaped depression of the gelatine, and the occlusion of it by an air-bubble during the first week or two, are however not invariably present; they are not present if the nutritive gelatine is weaker than 10 per cent. gelatine; in 2 to 7 per cent. gelatine the funnel-shaped depression and the occluding air-bubble is wanting, the liquefaction proceeding rapidly, and the surface end is from the outset marked by a drop of liquefied gelatine. The progress of the growth is considerably greater in such gelatine than in 10 per cent. gelatine. On inoculating from the same culture of the comma-bacilli two sets of tubes, one containing 5 per cent., the other 10 per cent. nutritive gelatine, and keeping them then under precisely the same conditions at 20° C., a marked difference will be found on inspecting the tubes after three to five days.

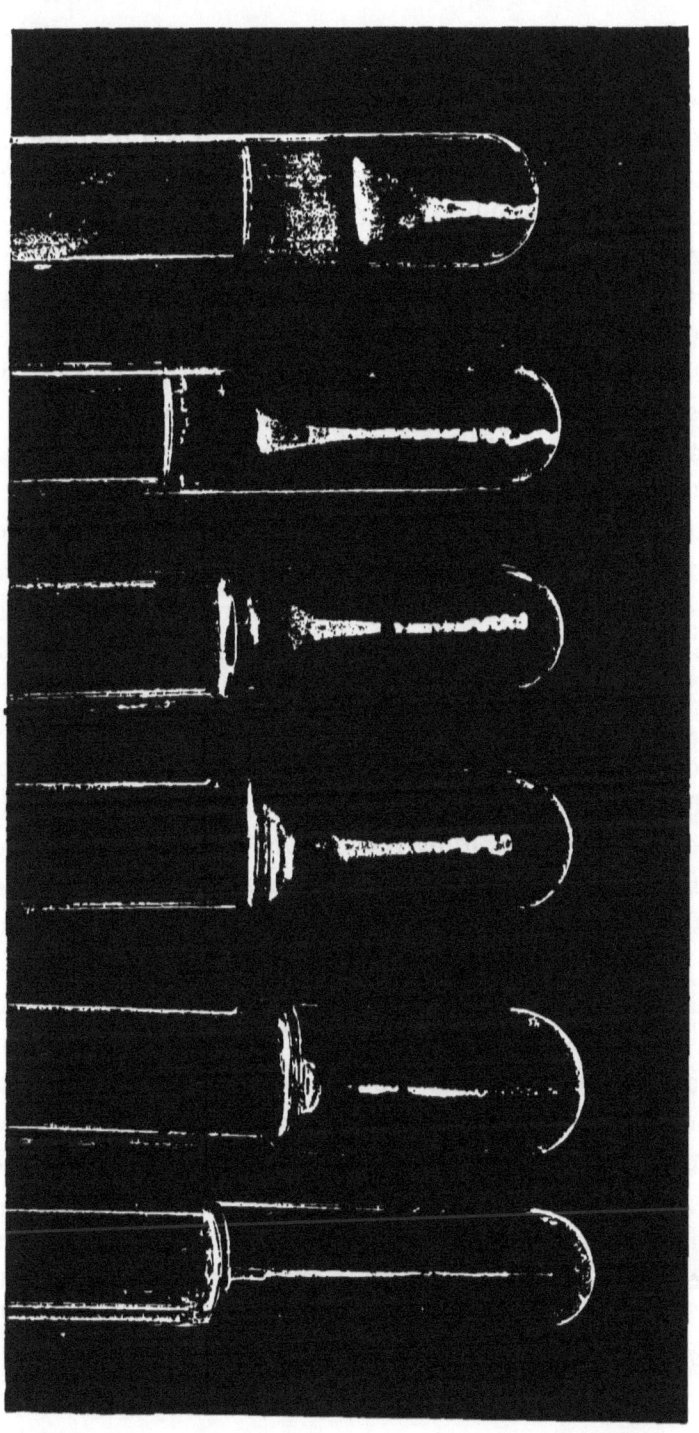

FIG. 26.—STABCULTURES OF CHOLERAIC COMMA-BACILLI IN NUTRITIVE GELATINE AFTER ONE, THREE, FOUR, FIVE, SEVEN, AND TEN DAYS RESPECTIVELY. COPIED FROM PLATE XII., "ARBEITEN AUS D. KAIS. GESUNDHEITSAMTE: ERFORSCHUNG DER CHOLERA."

Those containing 10 per cent. gelatine show the characters above described, the channel of inoculation is a thin greyish line of liquefied gelatine, funnel-shaped at the top, and with an occluding air-bubble, and at the bottom a small amount of precipitate; the other containing 5 per cent. nutritive gelatine, contains the growth in the shape of a broad cone of liquefied gelatine, broadest at the top, but with no funnel-shaped depression. The air-bubble is present in 10 per cent. nutritive gelatine, if during inoculation the comma-bacilli are well pushed down into the channel, as is sometimes the case when the inoculation is performed with a trace only, and by the platinum wire. With such a method of inoculation, most of the comma-bacilli are deposited at the bottom, few or none remain in the superficial parts, hence when active multiplication sets in, most of the growth takes place in the depth and away from the surface. By using a capillary glass pipette containing the comma-bacilli for inoculation, it can easily be arranged that the channel of inoculation receives comma-bacilli in its whole length. This can be achieved by holding the bulb of the pipette with the fingers, and thereby warming it, before withdrawing the pipette from the gelatine; not only is hereby a deposit of comma-bacilli ensured in the superficial layer also, but as a rule more comma-bacilli are deposited here than in the depth, because owing to the shape of the capillary pipette the channel of inoculation is at the outset broader on the surface than in the depth. When all these conditions are successfully fulfilled, it will be found that only a slight funnel-shaped depression will be noticeable after a few days, and consequently no air-bubble ever marks the mouth of the channel. The air-bubble, then, occluding the mouth of the funnel cannot be regarded as something quite typical and characteristic of a gelatine tube-cultivation of choleraic comma-bacilli, as many observers state, but is dependent on the strength of the gelatine, and particularly on the method of inoculation; no doubt in 10 per cent. nutritive gelatine, and under the same conditions of inoculation, *e.g.* by using a trace for inoculation and by the platinum wire, a conspicuous funnel-shaped depression of the surface and occlusion of it by an air-bubble will be noticed in most instances, but this is absent if the gelatine is weaker than 10 per cent., and if the inoculation is carried out in a different manner. Furthermore, I have seen very conspicuous funnel-shaped depressions and occluding air-bubbles in 10 per cent. nutritive gelatine tubes containing other than choleraic comma-bacilli; certain micrococci and several species of bacilli that possess the power of liquefying gelatine, when inoculated by means of a platinum wire, have in many instances shown the same funnel-shaped depression and the occluding air-bubble.

(*h*) *On potato.*—On boiled potato kept under a moist bell-glass the choleraic comma-bacilli grow readily when kept at temperatures varying between 32° and 37° C. From the place of inoculation the growth extends in the shape of a thin light-brown translucent film, in which, as time goes on, say after one to two weeks, thicker brownish spots and patches, due to local increase of the growth, appear. At temperatures varying between 18° and 22° C., even after several days, only a trace of growth can be made out with a lens.

CHAPTER V.

VARIOUS SPECIES OF COMMA-BACILLI.

WHEN, in 1883, Koch first announced from Egypt that in all cases of Asiatic cholera examined he had discovered in the dejecta and intestinal contents a species of bacilli which, "owing to their peculiar form, were called comma-bacilli," he was, according to his own showing,[1] not yet acquainted with their peculiarities in gelatine cultures; and after he had concluded his observations in Egypt, India, and France, in 1884, he stated[2] that in a large number of cases of intestinal disease the contents of the intestine had been examined, but "never was there found any trace of comma-bacilli." The intestinal discharges of dysentery and of infantile diarrhœa, the saliva of the mouth of various animals and the intestinal contents from various animals poisoned by arsenic, were searched, but no comma-bacilli were ever found. "Wherever I could get hold," he says,[3] "of a fluid containing bacteria, I examined it for comma-bacilli; but never have I found them. Only once I found in water from a salt-water lake in Calcutta a species of bacteria which at

[1] *Conferenz zur Erörterung d. Cholerafrage:* Berlin, July 26, 1884, p. 22.
[2] *Loc. cit.* p. 24.   [3] *Loc. cit.* p. 25

CH. V.] VARIOUS SPECIES OF COMMA-BACILLI. 83

first sight presented a certain resemblance to the cholera-bacilli, but on more careful examination they were somewhat thicker and did not liquefy nutritive gelatine." From this it is quite evident that, with the exception of this last instance, Koch had failed to find comma-bacilli anywhere except in cases of cholera. If Koch had known at that time that comma-bacilli occur in the saliva of the mouth and in various intestinal discharges, he would have no doubt added these instances to the one of the salt-water lake in Calcutta; but from his giving this as the only exception, there can be no question that he had failed to meet with

FIG. 27.—COVER-GLASS SPECIMEN OF FINKLER'S COMMA-BACILLI FROM A GELATINE CULTURE.
Magnifying power about 600.

comma-bacilli in any of the substances that were said to have been examined. He therefore felt justified, he thought,[1] in pronouncing that the comma-bacilli are constantly present in cholera Asiatica, but do not occur anywhere else. As is now well known, comma-bacilli are not so rare as Koch thought, but on the contrary are of rather common occurrence.

If in Egypt, while as yet unacquainted with the peculiar and distinguishing characters of the comma-bacilli in cultures in nutritive gelatine, and while, therefore, relying solely on

[1] *Loc. cit.* p. 25.

morphological appearances, Koch had found, as he might have done by more careful examination, that in the fluid of the mouth of man, in the intestinal contents of various intestinal disorders in man and animals, and in the normal contents of the intestines of some animals, comma-bacilli identical in morphological respects with the comma-bacilli in Asiatic cholera were constantly present, I am inclined to think that he would have adhered to his opinion formed before he went out to Egypt, viz., *that the distribution of the comma-bacilli in the intestine in cases of cholera* (sent him to Berlin from India some time previously), *proves them to be septic organisms*.

It is, I think, necessary to go back to this history of the discovery of the choleraic comma-bacilli, in order to show that the importance ascribed to them by Koch in relation to cholera may have had a good deal to do with his inability to find comma-bacilli anywhere else except in cholera cases. Koch says that since comma-bacilli occur only in cholera, and since he has failed to find them in the normal intestine, he concludes that the organisms and the disease stand in a direct relation. Now from his having failed to find comma-bacilli in a variety of localities where we now know that they constantly occur, his positive statement that they do not occur in the normal intestine loses a good deal of its value; for, as will be pointed out below, even if they did occur in very small numbers amongst the myriads and myriads of bacteria normally inhabiting the alimentary canal, it would be well-nigh impossible to demonstrate them; it would at any rate require long and exhaustive examination by many workers to establish their presence or absence on a reliable basis. If any one who has failed to find comma-bacilli in a variety of localities in which he has searched for them, but in which they have been shown by others to exist, tells us that in the normal human intestine no comma-bacilli occur, I think we shall be justified in showing some hesitation before accepting his dictum. What we do know of the characters, morphological and cultural, of the innumerable species of bacteria occurring in the different parts of the normal human intestine, is utterly insignificant as compared with what we do not know of them, and a great deal of work will have to be done still before any one is justified in saying that such and such a form present in a certain disorder of the intestine is not present in the normal state. Of course, if there are other evidences, such as the direct test of physiological action on the animal body, a conclusion can be drawn with certainty, but in the absence of such a test—and as I shall show this test is not forthcoming as regards the cholera-bacilli and cholera—it is quite unjustifiable for any one to pronounce so definitely as Koch did. Supposing that a particular pathological state of the ileum, particularly the mucus-flakes therein, were to favour the multiplication of the comma-bacilli already present in very small numbers in the normal state, then we should not be surprised to find that in cholera commabacilli are abnormally numerous (*see* v. Emmerich's statements in the *Archiv f. Hygiene*, Band III.).

Mr. Watson Cheyne, in the discussion on Asiatic cholera at the Medico-Chirurgical Society (March 1885), was thought by some present to have made a great point when he asked: How is it that if you want to demonstrate Koch's cholera-bacilli you have only to examine the intestinal contents of cholera cases, be they in India, in Egypt, France, or anywhere else, while you cannot find them in other cases of disease? No doubt, cholera being a disease peculiar to the human species only, and assuming that the cholera-intestine favours the multiplication of the comma-bacilli, it would follow that they are easily demonstrable in such cases; but that does not necessarily imply that they were not already present in very small numbers before the disease set in, and if so, of course their being found in India, Egypt, or France would make no difference to the conclusion so long as we have to deal with the human species.

I presume very few will dissent from the proposition that in many a putrid fluid crowded with all kinds of bacteria, it would be well-nigh impossible to discover *Saccharomyces*, although we know *Torula* is one of the most common organisms contaminating the air, and every fluid exposed to the air would receive and contain a good many examples of it.

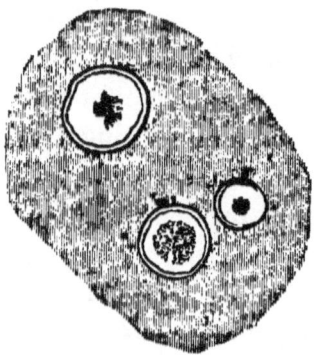

FIG. 28.—GELATINE PLATE-CULTIVATION OF FINKLER'S COMMA-BACILLI AFTER INCUBATION FOR FORTY-EIGHT HOURS AT 20° C.

But owing to the putrid fluid being an unfavourable soil, and particularly owing to the luxuriant growth in it of saprophytic bacteria, the comparatively few *Torulæ* at the outset present will not multiply. But transfer some of that fluid into a new medium containing besides traces of proteid material a good deal of sugar, and after a few days you will have no difficulty in showing the existence in this new medium of *Saccharomyces*. The conclusion to be drawn from this is obvious. Neither Koch, nor anybody else, has sufficiently and systematically examined the normal human intestine, and, as I have shown before, there is very little known about the nature and characters of the many species of bacteria present in the normal intestine: therefore the off-hand statement made by Koch and his adherents, viz. that comma-bacilli identical with those found in Asiatic cholera do not exist in the normal intestine is, to say the least, premature and not justified by their own limited observations.

In this connection, and as a confirmation of what has just now been said, I will quote Dr. Koch's own words, uttered in the discussion that followed[1] the reading of his paper. He says: " The question as to whether there exists any other disease, or any other condition in the human subject, wherein this same (comma) bacillus occurs, cannot be at present solved ; it will take years for its solution, and it will be necessary from time to time to examine in this direction any new disease that occurs. A strictly scientific decision (as to whether these same comma-bacilli belong exclusively to Asiatic cholera) is therefore at present impossible." In June 1884, a preparation of Koch's from the mucus-flakes of the ileum of a case of Asiatic cholera was shown in London to me and a number of others interested, by a gentleman who was indirectly associated with Koch in Egypt ; in this specimen the comma-bacilli were easily recognised. I mentioned on that occasion that I possessed specimens of the intestinal contents from an epidemic of bad diarrhœa that had occurred in Cornwall in the autumn of 1883, in which the same forms of comma-bacilli occur ; to this the gentleman answered with a smile and a shake of the head, so convinced was he from the teaching of Koch that comma-bacilli are present in Asiatic cholera exclusively.

I. In 1884 Finkler and Prior demonstrated and described

[1] *Loc. cit.* July 29, 1884, p. 55.

before the meeting of Naturalists and Physicians at Magdeburg, specimens of the intestinal dejecta of *cholera nostras* in which comma-bacilli occurred abundantly. They had cultivated them, and had thus obtained them in large numbers. True, the dejecta from which the comma-bacilli were obtained had been kept for some days, and the methods of cultivation employed by Finkler and Prior were not free from

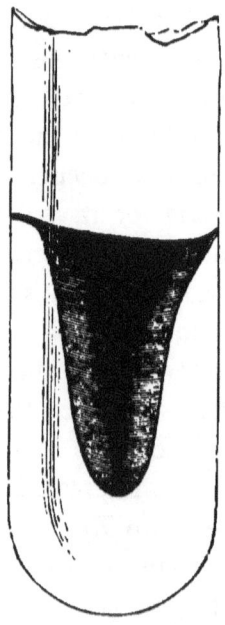

Fig. 29.—Cultivation of Finkler's Comma-bacilli in Nutritive Gelatine (10 per cent.) after four days' Incubation.

objection, and as a matter of fact their cultivations were at that time not pure cultivations of comma-bacilli; still, whatever might be urged against their description of the morphological changes they ascribed to those comma-bacilli, the fact remained and could not be criticised away, that veritable comma-bacilli possessing each and all of the morphological characters of the choleraic comma-bacilli—single commas more or less curved, and S-shaped and spiral forms—had been found elsewhere than in Asiatic cholera. Doubtless this was not at all a welcome discovery to Koch and his adherents, who had so frankly stated that notwithstanding their careful and exhaustive examination, they had "never seen any bacteria resembling comma-bacilli." The only thing to be done was to show that the comma-bacilli of Finkler and Prior were a different species from those occurring in Asiatic cholera.

Finkler and Prior then have proved the existence of comma-bacilli in a disease other than Asiatic cholera.

I have been fortunate enough to receive from them a mounted slide and a gelatine tube of the comma-bacilli, and there can be no question that while these comma-bacilli are in general respects similar to the choleraic comma-bacilli, they nevertheless present certain well-marked differences. The characters shown by these comma-bacilli are: (*a*) They occur as single commas, some more, others less curved, as S-shaped forms, and as wavy or more distinctly spiral forms; (*b*) they are possessed of motility, exactly like the choleraic comma-bacilli; (*c*) they show in well-stained and well-washed specimens the same distinction between sheath and protoplasmic contents, generally accumulated at the ends, as the choleraic comma-bacilli; (*d*) they are distinctly thicker and longer than the choleraic comma-bacilli; although this may not be striking when the comparison is made under a low power (say 300–400), it is conspicuous when a specimen of choleraic comma-bacilli made from a gelatine culture is compared under a high power with one of Finkler and Prior's comma-bacilli grown in a like medium; (*e*) cultivated on Agar-agar mixture, in broth, or vegetable albumen and Agar-agar, and on egg-albumen and Agar-agar, the characters of the growth are very much the same as those of the choleraic comma-bacilli—they grow rapidly and well at temperatures of from 30°—37° C.; (*f*) on boiled potato Finkler's comma-bacilli grow well even at temperatures of 18°-22° C., and from the spot of inoculation a smeary greyish-brown film soon expands, and gradually thickens; in a few days at 20° C., the growth is copious; in this respect then a marked difference exists between the two comma-bacilli; (*g*) in nutritive gelatine in plate-cultivation the character and aspect of the colonies of Finkler's comma-bacilli is very similar to those of the choleraic comma-bacilli, except that the colonies make their appearance much sooner, and having appeared, grow much more rapidly than those of the choleraic comma-bacilli; (*h*) they liquefy the gelatine, but the liquefied gelatine both in plate-cultivation and in tubes is less clear than is the case with the choleraic comma-bacilli; on comparing a gelatine tube of Finkler's comma-bacillus with one of the choleraic comma-bacilli, after a week or more the liquefied gelatine in the latter is more clear than in the former, although in both it is not quite clear; gelatine tubes in which Finkler's comma-bacillus has been growing, and in which the gelatine with the exception of the deepest parts has become completely liquefied, show the same whitish or greyish-brown precipitate as similarly advanced culture-tubes of choleraic comma-bacilli, the liquefied gelatine being only slightly opaque and in the upper layers almost translucent, while in similarly advanced culture-tubes of choleraic bacilli the liquefied gelatine is a little less opaque, and in many tubes there is present on the surface a granular whitish filmy pellicle (p. 78); (*i*) the best test of distinction is no doubt that pointed out by Koch, namely, the mode of growth and the rapidity with which the two kinds of comma-bacilli grow in nutritive gelatine of 10 per cent.

v.] VARIOUS SPECIES OF COMMA-BACILLI. 91

strength. The appearance of a series of gelatine-tubes inoculated by means of the platinum wire or capillary glass pipette with Finkler's comma-bacillus, and of another series of similar tubes inoculated in the same manner with choleraic comma-bacilli and kept at 20° C. from three to four days, leaves no doubt that they contain two different species. While the cultures of the choleraic comma-bacilli are only in their early stage, showing the channel of inoculation as a thin greyish line of liquefied gelatine with a trace of whitish precipitate at the bottom, and a distinct funnel-

FIG. 30.—COVER-GLASS SPECIMEN OF MUCUS-FLAKES FROM A MONKEY SUFFERING FROM DIARRHŒA.
1. Straight bacilli containing bright oval spores.
Magnifying power about 600.

shaped depression of the surface with an occluding air-bubble, those of Finkler's comma-bacillus show a broad conical growth; the liquefied gelatine occupying almost one-third of the breadth of the tube, and being uniformly turbid. In the earlier stages, say after thirty to sixty hours, the surface shows the funnel-shaped depression as well as the air-bubble, but after three days as a general rule the liquefaction has so far progressed that of the funnel-shaped depression and occluding air-bubble little is left.

A curious fact which I have repeatedly observed is

this : if, instead of 10 per cent. gelatine, nutritive gelatine of the strength of 3 to 6 per cent. is substituted, the mode and rapidity of growth of the two kinds of comma-bacilli are indistinguishable, and it seems to me probable that Finkler and Prior's statement that the two kinds of comma-bacilli do not differ in their mode and rapidity of growth in nutritive gelatine must be due to their having used gelatine of less strength than 10 per cent. I am quite sure that Koch is right in saying that as a general rule 10 per cent. nutritive gelatine shows the well-marked differences mentioned above. But I do not think Koch is right when he says that while the growth in tubes of choleraic comma-bacilli possesses an aromatic smell, those of Finkler's have a putrid smell ; for there is little difference noticeable in this respect between the two, except that in the latter there is certainly not that distinct aromatic smell as in the former.

The question whether Finkler's comma-bacilli stand in any definite relation to *cholera nostras*, as is maintained by Finkler and Prior,[1] has been definitely set at rest by the following considerations :—(1) The observations made by Koch and Frank have failed to demonstrate the presence of these comma-bacilli in typical cases of *cholera nostras*, therefore their number cannot at all events be remarkably great in this malady ; (2) Dr. Miller of Berlin has proved by cultivation that comma-bacilli apparently identical with those of Finkler and Prior occur in the mouth in connection with caries of the teeth ;[2] (3) Kuisl succeeded in isolating by cultivation from normal human fæcal matter comma-bacilli which morphologically and in culture are identical with Finkler and Prior's comma-bacilli.[3]

[1] *Ergänzungshefte zum Centralblatte f. allgem. Gesundheitspflege*, vol. i. parts 5 and 6.
[2] These bacilli of Miller are however not considered quite identical with Finkler's.   [3] *Aerztl. Intelligenzblatt*, 36 and 37, 1885.

VARIOUS SPECIES OF COMMA-BACILLI.

II. The late Dr. T. R. Lewis was the first to point out[1] the occurrence, in the fluid of the mouth, of comma-bacilli morphologically identical with those of Asiatic cholera. They are of the same size, of the same kind of curvature, single or S-shaped, and they move like the choleraic comma-bacilli. This simple fact, which is easily verified, had

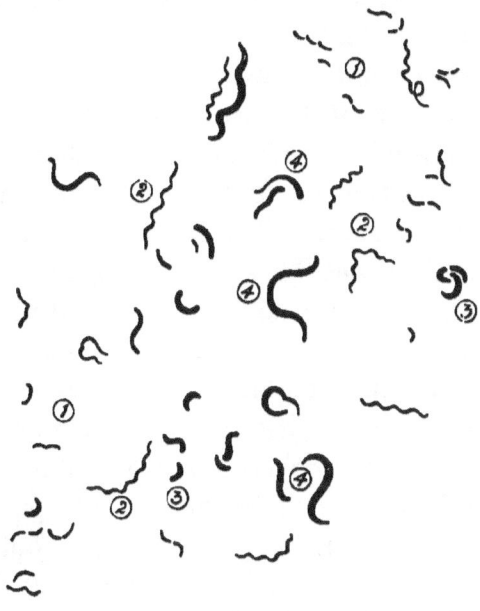

FIG. 31.—COVER-GLASS SPECIMEN OF CONTENTS OF CÆCUM FROM A NORMAL GUINEA-PIG.
1. Small comma-bacilli.
2. Spirilla of same.
3. Larger comma-bacilli.
4. Large coiled organisms.
Magnifying power about 700.

entirely escaped Koch. He said[2] that he had examined the saliva of the mouth and had not seen anything like comma-bacilli. It seems, therefore, to say the least, very remarkable that some of Koch's adherents should have made so light of Lewis's discovery; they appeared to take

[1] *Lancet*, September 20, 1884.   [2] *Loc. cit.* p. 25.

this occurrence of comma-bacilli as a matter of course, which "every one knew before." It must be plain to every impartial reader that not even Koch knew this until Lewis pointed it out.

The comma-bacilli of Lewis are not the same as those afterwards described and isolated by Miller, since the latter have been isolated and grown in 10 per cent. alkaline gelatine, and behave similar to those of Finkler and Prior, while those of Lewis do not behave in this way. Neither Lewis nor Koch, nor any of the many workers in Koch's laboratory who have tried to grow them in alkaline 10 per cent. gelatine, have succeeded. It occurred to me that inasmuch as the fluid of the mouth often has a neutral or even faintly acid reaction, it might be possible to grow these comma-bacilli in neutral or faintly acid nutritive gelatine. I examined the fluid of my own mouth, and as had been already shown by Lewis, I found that comma-bacilli vary greatly in numbers at different times: sometimes I could find in every specimen (made by drying on a cover-glass a thin film, and staining it afterwards in gentian-violet or Spiller's purple) several examples of comma-bacilli; at other times only in one or the other could a few be found, while at other times again I found numbers of them in groups and as isolated examples. On such occasions I made a number of plate-cultivations in 5 per cent. neutral nutritive gelatine, and in one instance after several fruitless attempts I did get a colony of liquefying comma-bacilli, which in their manner of growth in the plate-cultivation seemed indistinguishable from the choleraic comma-bacilli; from such a colony 10 per cent. alkaline nutritive gelatine in test-tubes was then inoculated, and the growths produced therein were not distinguishable from those of choleraic comma-bacilli. I have afterwards, on many occasions, repeated the original experiment, and have had, after a great many failures, one other successful colony produced in plate-cultivation in neutral nutritive gelatine from which a series of tubes of 10 per cent. alkaline nutritive gelatine were started. These I have kept growing for many generations, and their behaviour in Agar-agar mixture, in broth, in gelatine, and in potato was carefully noted and compared with cultures of choleraic comma-bacilli; all I can say is that they appear to me identical. The only difference that I can find is that the comma-bacilli from the colony in the plate-cultivation appeared slightly larger, that is to say, thicker than the choleraic comma-bacilli, but this difference was not so striking in the cultivations in 10 per cent. alkaline gelatine.

I maintain then that in the normal fluid of the mouth there occur at least two kinds of comma-bacilli, one very similar to the choleraic comma-bacillus, the other (isolated by Miller) similar to Finkler's.

III. Deneke described comma-bacilli which he found in stale cheese, and which he afterwards isolated and cultivated. Morphologically, and also in gelatine cultures, they appear almost identical with those of Asiatic cholera, so much so that he failed to observe any describable difference; in fact, he states that so far as he was able to observe, the only distinction seems to be their different action when injected into guinea-pigs. He afterwards, however, modified this view, inasmuch as Flügge showed that there are slight differences from the choleraic comma-bacilli in their mode of growth in nutritive gelatine, in which respect they stand about midway between the choleraic and Finkler's comma-bacilli. At any rate this much seems certain, that the differences existing between them and Koch's comma-bacilli are not very great, not so great, in fact, as those between

Finkler's and Koch's comma-bacilli. On potato they do not grow at all. I have myself had an opportunity, thanks to Dr. Crookshank, of examining and cultivating these cheese-comma-bacilli or cheese-spirilla, and am able to say that in gelatine cultures they are difficult to distinguish from the choleraic comma-bacilli.

IV. As has been already stated, Koch in his first pamphlet embodying the results of his investigations on cholera in Egypt, India, and France, showed that he was then unaware of the existence of comma-bacilli other than those in Asiatic cholera, *with the exception of one instance*—that of a salt-water lake in Calcutta, in which he found comma-bacilli that looked like choleraic comma-bacilli, but did not behave like them in cultivation—he had never seen any bacteria that looked like the comma-bacilli. This being the case, our finding comma-bacilli in the intestinal contents of cases of diarrhœa, dysentery, and phthisis seemed not without interest. Hence the strong adverse criticism expressed by Mr. Watson Cheyne during the discussion on cholera at the Medico-Chirurgical Society in 1885, at my having mentioned such a fact without saying how these comma-bacilli behave under cultivation in gelatine, was not quite justified. The statement made by me in the preliminary Report of the English Cholera Commission, that comma-bacilli do occur in intestinal diseases other than Asiatic cholera, was in relation to Koch's statement that in no other disease of the alimentary canal, nor in any other circumstance—except the above-quoted salt-water lake in Calcutta—had he ever seen bacteria that looked like the comma-bacilli. The comma-bacilli which we saw in the intestinal contents of cases of diarrhœa, dysentery, and phthisis, looked thicker and longer than those in cholera, and their mode of growth in gelatine was not then ascertained. But I have since my return to England ascertained that comma-bacilli obtained from a case of severe diarrhœa in an adult do grow in 10 per cent. gelatine; in weaker gelatine (5 per cent.) they grow well, but their mode of growth is no doubt altogether different from that of the choleraic comma-bacilli; although they liquefy the gelatine, they do so very rapidly, and the liquefied gelatine is turbid, thick, and smells offensively. A very striking case of the occurrence of crowds of comma-bacilli in the intestine in diarrhœa I have met with in a monkey.

Fig. 32.—Cover-glass Specimen from a Cultivation in 10 per cent. Nutritive Gelatine of the non-liquefying variety of Comma-bacilli from a case of Noma in a child.
    1. Single comma-bacilli.
    2. Spiral forms.
    3. Wavy forms.
    Magnifying power about 700.

These animals in the summer months are very often affected with diarrhœa. We have had during several years at the Brown Institution several cases of severe diarrhœa in monkeys; some died, others recovered under treatment. One animal was killed thirty hours after diarrhœa set in. The stools were liquid, yellow, and contained granules of fæcal matter; in the large intestine was a quantity of the same yellow thin fluid, in it were suspended numerous mucus-flakes. I may state that this animal had been purchased in good health about a fortnight before, and had been kept, like the other monkeys, in a separate stall, uniformly warm and well-ventilated. The food was copious and of the ordinary kind—potatoes, rice, and milk.

Preparations made of the mucus-flakes of the cæcum revealed, besides straight thick spore-bearing bacilli, slightly pointed at the ends, large numbers of motile comma-bacilli. In Fig. 30 I have given an accurate representation of a number of these comma-bacilli present in the same place in the mucus-flake. The identity in morphological appearances and in size with the choleraic comma-bacilli is very striking indeed; there are the same single commas, and S-shaped, circular, and semi-circular forms. I possess a good many preparations (stained and mounted) of the mucus-flakes of the ileum of cases of acute typical cholera in which the number of comma-bacilli is not by any means so great as in the case of this monkey. I have made a number of plate-cultivations from the contents of the cæcum of the monkey, but owing to the hot weather then prevailing (June) the above-mentioned straight bacilli had in the course of twenty-four hours so increased in numbers and so pervaded the gelatine that the whole became liquefied and crowded with them. Owing to this the cultivations became altogether useless. In another monkey that died with diarrhœa, the contents of the cæcum were also crowded with comma-bacilli, many of these S-shaped and spiral; but these were conspicuously thicker than those in the former case.

Of other cases of comma-bacilli in the contents of the cæcum in monkeys I shall have something to say later on.

V. In his second paper[1] Koch says he has repeated, with positive results, the experiments made first by Nicati and Rietsch on guinea-pigs (namely, injection of cultures of

[1] *Deutsche med. Woch.* 45, 1884.

VARIOUS SPECIES OF COMMA-BACILLI.

choleraic comma-bacilli direct into the duodenum); and in his third paper [1] he gives the results of a series of successful experiments on guinea-pigs. In both instances mention is made of the multiplication of the injected choleraic comma-bacilli in the intestines of the guinea-pigs experimented upon, but I find no evidence that he had first sought for the presence of comma-bacilli in the contents of the intestines of normal guinea-pigs.

Fig. 33.—Plate-cultivation of the same non-liquefying Comma-bacilli of Noma as in Fig. 32.

The plate-cultivation is several weeks old. The drawing represents the colonies as seen under a lens. The colonies marked by uniform shaping are situated in the depth.

Van Ermengem [2] gives a figure (Pl. XI. Fig. 1) of comma-bacilli in the intestinal contents of a guinea-pig that died after intraduodenal inoculation, and he describes them (p. 372) as "*grandes virgules fortement incurvées et d'aspect assez anormal.*" He tells us [3] that he found comma-bacilli in the large intestine of guinea-pigs, but they are much

---

[1] *Conferenz zur Erörterung der Cholerafrage:* Berlin, May 1885.
[2] *Recherches sur le microbe du Choléra Asiatique.*
[3] *Loc. cit.* p. 87.

larger than Koch's and their cultures in gelatine plates are altogether different. While I have no doubt that his *grandes virgules fortement incurvées* correspond not to Koch's comma-bacilli, but to certain forms normally present, I have grave doubts about his colonies of these normal comma-bacilli from the guinea-pig. I have tried over and over again to isolate these by plate-cultivation, but have never succeeded in growing them.

A somewhat similar statement is made by Mr. Watson Cheyne on p. 13 of his pamphlet (a reprint of a series of articles that had appeared in the *British Medical Journal* of April 25, May 2, May 16, and May 23, 1885): "One of the most peculiar forms (of comma-bacilli) which I have seen was found in the contents of the large intestine of guinea-pigs, which died after injection of cholera-bacilli. I tested the fluid by cultivation at the time very carefully, and found that it contained almost a pure cultivation of cholera-bacilli; there was certainly not more than one other kind of bacilli for every hundred cholera-bacilli. The appearance of this material, on microscopical examination, after staining, is shown in the accompanying figure (Fig. 5). Large, fat, coiled, almost worm-like organisms will be seen, which, as I know by cultivation, are cholera-bacilli, but which could not be recognised by the microscope alone." Now it is a fact, easily verified, that what Mr. Watson Cheyne here describes and figures, are forms present in the contents of the large intestine of every normal guinea-pig, as I pointed out in a note in the *British Medical Journal* (May 9, 1885).

Those organisms figured and described by Mr. Watson Cheyne, which he thought he identified by cultivation as cholera-bacilli, can be easily demonstrated by spreading on a cover-glass a thin film of the contents of the cæcum of any normal guinea-pig, previously diluted slightly with salt-solution, and then drying and staining in the usual manner. Crowds of these large, fat, coiled organisms seen by Mr. Watson Cheyne are met with, together with other smaller ones that look very much like choleraic comma-bacilli. It is, however, only fair to state that Mr. Watson Cheyne, becoming no doubt aware of his error, has afterwards inserted on p. 28 the following corrective statement:—"I think it most probable that these bodies [shown in his Fig. 5] are the cholera-bacilli which were found on cultivation to be present in enormous numbers, because there were no other markedly curved organisms present, and because they seemed to show all gradations between small slightly-curved rods, and the large, coiled bodies shown in the drawing." Now, this is exactly what is the case in the material taken from the cæcum of every normal guinea-pig that I have examined; small comma-bacilli, single and S-shaped and spiral, of the size and appearance of the choleraic comma-bacilli, and the large, fat, coiled organisms mentioned above, and all intermediate gradations.

In Fig. 31, I have given an accurate representation of the appearances in a specimen, prepared, stained, and mounted, of the contents of the cæcum of a normal guinea-pig, and it is evident from this that Mr. Watson Cheyne's former and later descriptions are equally applicable to this preparation. The above-mentioned large, fat, curved, and coiled organisms seem to me more like flagellate infusoria than comma-bacilli.

Messrs. Cornil & Babes have recently[1] given a drawing and description of the comma-bacilli present normally in the intestine of guinea-pigs. Their drawing does not exactly represent the appearances, and they do not state, as they

[1] *Microbes*, Second Edition.

ought to have done, that they are not the discoverers of these comma-bacilli in the intestine of normal guinea-pigs.

I have repeatedly tried to grow these normal comma-bacilli in 10 per cent. alkaline nutritive gelatine in plate-cultivations, but have not succeeded: there was not

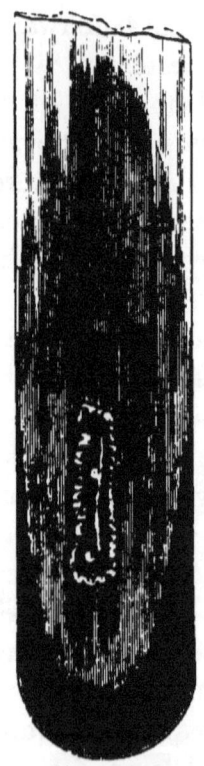 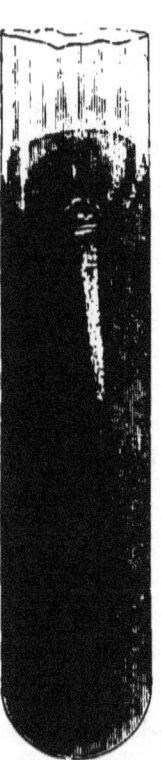

FIG. 34 and 35.—SAME NON-LIQUEFYING COMMA-BACILLI GROWING IN 10 PER CENT. NUTRITIVE GELATINE; SEVERAL WEEKS OLD.

FIG. 34.—On the surface.
FIG. 35.—After inoculation by stabbing.

even an attempt at growth, only straight bacilli and micrococci were thus obtained from the intestinal contents. Of comma-bacilli or spirilla forms, shown in Fig. 31, there was never any trace. Mr. Watson Cheyne, in his plate-cultivations made from the guinea-pig that had been infected *per duodenum* with choleraic comma-bacilli, obtained numerous colonies of the true choleraic comma-bacilli; this is of course proof that the choleraic comma-bacilli injected were still present and had multiplied in the intestinal contents.

Nicati and van Ermengem have also found comma-bacilli in the intestinal contents of the pig, rabbit, horse, and other animals, but they are said to differ from Koch's comma-bacilli.

VI. Comma-bacilli, in morphological respects indentical with the choleraic comma-bacilli, have been found by my colleague, Mr. Alfred Lingard, in a case of noma in a child. A noma, including the whole thickness of the mucous membrane of the lip of the mouth, was excised by the house-surgeon at University College Hospital for Mr. Lingard. On removal, cultivations were made from the depth of the tissue by means of the platinum-wire, and from the fresh tissues sections were cut in the ordinary way, stained and mounted.

In the sections there were found at the point of demarcation between the healthy (or rather inflamed) and the necrotic tissue, but situated in the former, numbers of comma-bacilli pervading the tissue in clumps and streaks. The cultivations in Agar-agar proved to be growths of the same comma-bacilli, but in an impure state. I then, by Koch's method of gelatine plate-cultivation, isolated the comma-bacilli and found that they belonged to two well-defined species, one smaller and one slightly larger variety. The smaller variety behaved in plate-cultivations, and in the cultivations in gelatine tubes established from these, very much like the choleraic comma-bacilli; the slowness of the colonies' growth, the manner in which they liquefied the gelatine, the aspect and nature of the colonies, the manner in which they grew in gelatine tubes, and the microscopic appearances of fresh specimens and of dried and stained specimens, were similar. The other and slightly larger variety behaved altogether in a different way. This did not liquefy the gelatine, it formed in plate-cultivations single greyish rounded specks (see Fig. 33) which only slowly enlarged, and even after a week or ten days were not larger than a few millimetres in diameter, their outline being irregular, their colour greyish, and their aspect under a lens more or less granular. In gelatine tubes (10 per cent. alkaline gelatine) inoculated by the platinum-wire in the depth—stab-culture, there appeared after several (three or four) days the first traces of the growth in the shape of a greyish streak, slowly thickening and broadening, and assuming a more or less granular aspect; on growing it on the surface of 10 per cent. alkaline nutritive gelatine in streak culture, a greyish film appears, gradually expanding in breadth, but even after several weeks it is only a few millimetres in breadth; in the streak of inoculation the growth is thickest, and assumes after two or three weeks a slight reddish-brown tint, while the rest is greyish and thin. After some months the film is about half an inch broad, thin and grey in the peripheral part, thicker and of a reddish-brown tint in the middle. The outline of the film is smooth in some places, slightly crenate in others. At no time, not even after some months, is there a trace of liquefaction of the gelatine. These comma-bacilli grow well on alkaline Agar-agar mixture, also, but not so copiously, on vegetable albumen and Agar-agar mixture, or egg-albumen and Agar-agar mixture, forming on all these media a greyish film, which as time goes on assumes in its thickest parts a brownish tint. In microscopic specimens made from gelatine tubes several weeks old there are found single commas and S-shaped forms, and wavy and spiral organisms. The latter are very interesting, inasmuch as many are made up of several very closely-twisted turns in the middle, while the ends are only slightly wavy. Examples of these forms are shown in Fig. 32.

VII. Dr. Emil Weibel has during the last two years described a number of different species of comma-bacilli (*Centralblatt für Bacteriologie und Parasitenkunde*, II. Bd., No. 16., IV. Bd., Nos. 8, 9, 10).

Von Emmerich (*Archiv für Hygiene*, Bd. III., p. 358) had already drawn attention to the prevalent occurrence of comma-bacilli or vibriones in various substances rich in mucus (the mucous fæces of helix) and he thought that probably owing to the presence of the mucus-flakes in cholera-Asiatica the comma-bacilli multiply so readily in this disease.

Weibel has followed this up and has isolated and cultivated comma-bacilli from the nasal mucus. He describes and figures this vibrio and found that it does not liquefy the gelatine. Similarly he describes and figures two species of comma-bacilli, which he isolated and cultivated in pure cultures made of putrid hay infusion, vibrio saprophyles α and β; neither of them liquefy the gelatine.

Another species which did not liquefy gelatine was isolated by Weibel from the mucus of the tongue; further he isolated a vibrio saprophyles γ from mud of drains, and finally he isolated three species whose growths are conspicuous by their yellow colour: vibrio aureus, vibrio flavus, and vibrio flavescens.

Gamaleïa (*Annales de l'Institut Pasteur*, 1888, No. 9, p. 482) states that an acute fatal disease affecting fowls in Odessa during the summer months, similar to, but not identical with Fowl-cholera, is caused by a species of comma-bacilli, which he calls "Vibrio Metschnikovi." This vibrio is said to be indistinguishable in morphological and

 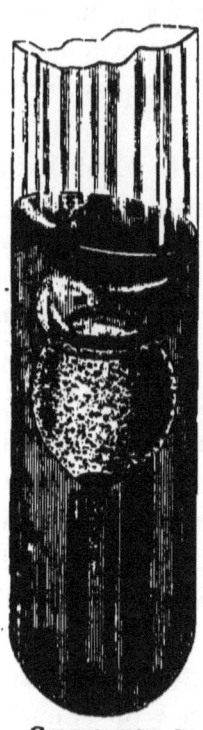

Fig. 36.—Cultivation in ten per cent. Nutritive Gelatine of a Micrococcus isolated from the blood of the Finger of a Person affected with Scarlatina.

The cultivation is one week old, and very much resembles a cultivation of choleraic bacilli.

Fig. 37.—Cultivation in Gelatine (10 per cent.) of same Micrococcus after three weeks, showing a large funnel-shaped Opening on Surface with an occluding Air-Bubble.

The main part of the growth is liquefied gelatine with numerous granules at the side and within it; the depth is not liquefied.

cultural characters from Koch's comma-bacilli of human Asiatic cholera.

We see then that the number of the different known species of comma-bacilli is already considerable and is constantly increasing.

As has been stated in the foregoing pages, the comma-bacilli found by Koch in Asiatic cholera have certain definite characters in culture media (particularly in nutritive gelatine) by which they can be easily distinguished from other species, although not from all. Certain other characters at first attributed to them are not exclusive.

In his first publications Koch gave us to understand that a notable character of the choleraic comma-bacilli is that they require for good growth an alkaline medium. Now this of course cannot and does not mean that they alone require for good growth an alkaline medium, or that they do not grow well in any but an alkaline medium. In the first place, all the different species of comma-bacilli described in previous pages, and other species of bacteria also (various micrococci and bacilli), grow very luxuriantly in an alkaline medium, some much more luxuriantly in alkaline than in neutral or faintly acid media. In the second place the comma-bacilli of Asiatic cholera live and grow well also in neutral and even faintly acid media. Thus the rice-water stools and the intestinal contents have been in several instances tested as to their reaction, and have been found of neutral, or when kept for several hours, even of faintly acid reaction, and nevertheless there were present in them numerous comma-bacilli in active motion and multiplying rapidly. I have repeatedly grown the choleraic comma-bacilli in neutral nutritive gelatine, in neutral Agar-agar mixture, and in neutral broth, and have thus obtained good and active cultures. Koch has shown that notwithstanding that the reaction of potato is acid, the comma-bacilli of cholera grow well on it.[1] Culture media (gelatine broth, broth peptone), though at starting faintly alkaline, become, when choleraic comma-bacilli grow in them, faintly but

[1] *Loc. cit.* pp. 18, 19, and 20.

distinctly acid, even after a few days only, and long before the climax of growth is reached. I have in one instance convinced myself that they are capable of growing (though of course not copiously, any more than in the case of many other bacteria), even when the nutritive medium (broth) was at the outset just faintly acid.

Another assertion made by Koch was that the comma-bacilli are killed by acid, provided this be of the requisite strength. While there can be no question after the detailed experiments of Mr. Watson Cheyne that hydrochloric acid of the strength of 0·2 per cent. kills without fail the comma-bacilli, if allowed to act on them for a few minutes, I do not hesitate to say that with the exception of the spores of bacilli all other bacteria—micrococci, bacteria, sporeless bacilli, and vibriones and spirilla—are alike killed under similar circumstances by acid of such strength.[1] And just as a more dilute acid does not affect in a few minutes' action the vitality of many other bacteria, so also the comma-bacilli remain unaltered by it. I have mixed choleraic comma-bacilli from a pure culture with a mixture of one part of commercial hydrochloric acid in 1,000 parts of water for from ten to thirty minutes, and on inoculating broth with the medicated comma-bacilli I have obtained normal growths of them. Other media known to have a noxious effect on bacteria in general, have the same effect on the choleraic comma-bacilli. The experiments of Koch and also of Mr. Watson Cheyne ought, however, to be accepted with a certain restriction after Dr. Macfadyan's experiments carried out on the living animal.[2] Dr. Macfadyan has shown that if a dog be kept without food

---

[1] Compare also a paper by Mr. Laws in the *Report of the Medical Officer of the Local Government Board*, 1884.

[2] *Journal of Anatomy and Physiology*, vol. 21, 1887.

for twenty-four hours, and then water containing the choleraic comma-bacilli be introduced into the stomach, the comma-bacilli can be recovered by cultivation from the contents of the jejunum and ileum four hours after, proving that the comma-bacilli have passed the stomach in a living state. If, however, they are given to the animal in milk under the same condition, they cannot be recovered by cultivation from the small intestine.

Koch in his first memoir has given us the results of numerous systematic experiments made with regard to the influence on the growth of the comma-bacilli of the absence and presence of oxygen, with regard to the influence of sufficient or deficient nutriment, and especially with certain substances, such as alcohol, iodine, carbolic acid, cupric sulphate, quinine, corrosive sublimate, &c. Van Ermengem in his book states that he has repeated these experiments.

Hueppe has grown the choleraic comma-bacilli within the hen's egg and has found that they, as well as the comma-bacilli of Finkler, as also those of Deneke, are capable of growing luxuriantly under these conditions, as also when growing after Reichert's method, *i.e.* exclusion of all free oxygen, thus showing that the comma-bacilli can grow well anaerobically. One striking result of Hueppe's experiments was the demonstration that all these three species of comma-bacilli form sulphureted hydrogen when grown in egg.

Another remarkable fact to be again referred to further below was the production in egg by the three kinds of comma-bacilli (Koch, Finkler, Deneke,) of a highly toxic chemical virus, far more distinct and striking than what is taking place in gelatine cultures or broth cultures (see below.)

I have myself made some experiments on the choleraic comma-bacilli and other bacteria with phenyl-propionic and phenyl-acetic acid, with perchloride of mercury and with iodate of calcium. A noteworthy result of these experiments (published in the *Report of the Medical Officer of the Local Government Board* for 1885) was this—that while the killing and restraining power of corrosive sublimate on the choleraic comma-bacilli, which by the way is greatly inferior [1] to what was found by Koch, is very much the same as on Finkler's comma-bacilli, and some notoriously saprophytic bacteria, it is not so great on them as on some notoriously pathogenic bacteria (exclusive of spore-bearing forms).

Strong solutions of iodate of calcium, while possessed of antiseptic action on micrococci, as well as on pathogenic sporeless bacilli, have no effect on the choleraic comma-bacilli; when kept mixed with the solution the comma-bacilli retain their motility and their power of multiplying unimpaired. Sea-water has likewise no disinfecting action on them.

With regard to the aspect and character of the colonies in gelatine plate cultivations, although they are well-marked in the case of choleraic comma-bacilli (see p. 71), they are not exclusively confined to them; I have seen colonies of micrococci (obtained accidentally from necrotic tissue, and from urine) which very much resembled them in general aspect, and in the mode of liquefaction of the gelatine, but they grew more rapidly.

Finally as to the character of the growth in gelatine tubes, particularly the funnel-shaped depression of the surface and the occluding air-bubble, I have observed these in the case of some other bacteria, as has been already stated. But the general appearances as growth proceeded became somewhat different from those presented by the choleraic comma-bacilli.

Dr. Odo Bujwid pointed out [2] that the cultures of choleraic

[1] Koch gives the restraining power of perchloride of mercury on choleraic comma-bacilli as measured by 1 in 100,000 of water; I find that not even 1 in 30,000 has such a power.

[2] *Zeitschrift für Hygiene*, Bd. II. 1, p. 52.

comma-bacilli in broth at 37° C. show already after twelve hours' incubation, on the addition of the amount of 5 to 10 per cent. of common hydrochloric acid (also nitric or sulphuric acid) a characteristic pink colouration which rapidly increases in intensity during the first half hour; it lasts for a few days, and exposed to light changes into a brownish tint.

Dr. E. K. Dunham has shown[1] that for this reaction the presence of peptone is essential, and that the pink colour becomes more pronounced and more rapidly evident if instead of HCl, concentrated sulphuric acid be employed. A previous addition of a drop of nitric acid enhances the reaction, which under this condition can be obtained also with Finkler's and Deneke's comma-bacilli. Salkowski had previously shown that this reaction is due to the formation of indol and a trace of nitrite.

Zeitschr. für Hygiene, II. 2, p. 337.

CHAPTER VI.

DIAGNOSTIC VALUE OF CHOLERAIC COMMA-BACILLI.

So far we have seen that various species of comma-bacilli are known, and that of these the choleraic comma-bacilli possess certain definite characters in cultivations on nutritive gelatine which are not possessed either by those of Finkler and Prior, Miller, Kuisl, Deneke, Weibel and others, by one form of those observed in noma, or by those which I observed in the diarrhœa of man, and in the contents of the cæcum of the guinea-pig. Those which I observed in the contents of the cæcum of the monkey suffering from diarrhœa have not yet been cultivated. But those which occur in the fluid of the mouth, and which are probably those observed by T. R. Lewis, I think I have in two instances out of many succeeded in cultivating, and they appear to me to be in their manner of growth in nutritive gelatine strikingly similar to the choleraic comma-bacilli. Lastly, one of the two species observed in noma was found to grow in gelatine in the same manner as the choleraic comma-bacilli. Now it has been often stated, and is by many held, that two kinds of organisms, morphologically alike, and growing in a like manner in the various artificial media commonly in use, must of necessity be one and the same species. There would be no more

CH. VI.] VALUE OF CHOLERAIC COMMA-BACILLI. 113

justification for a general statement of this kind than if one were to lay it down that two species notoriously different in chemical and physiological respects must always show different characters in the artificial cultivations commonly employed; for the contrary has been proved as regards several species. It has for instance been shown that the capsulated oval coccus first discovered by Friedländer in connection with croupous pneumonia is not the only species that in morphological respects and in artificial cultures possesses the characters he described. Of other bacteria, *e.g.* streptococcus (some of those studied by Rosenbach, by Uffendi and others), the same holds good.

Comparing for instance the behaviour of some of the non-pathogenic micrococci studied by these last two observers in artificial cultures with that of some of the pathogenic micrococci (*e.g.* of erysipelas and of vaccinia) the similarity is very striking; nevertheless the physiological action is totally different. The same is to be said of the bacillus of typhoid fever, and of some of the species of the proteus of Hauser. If two kinds of bacteria differ in mode of growth, *cateris paribus*, in the same culture media, I think it will be admitted that they are probably different species. While then it will be admitted that Koch's comma-bacilli must be of a different species from those of Finkler and Prior, of Miller, and of the species described above as found in noma, it does not follow that they are identical with those cultivated by Deneke or by myself from the mouth, from a case of noma, and (as I shall hereafter mention) from a monkey that had been subjected to certain definite experiments. And it seems to me that Koch and his adherents go too far in trying to make out first, that the comma-bacilli belong exclusively to Asiatic cholera, and then, after this had been disproved, that they differ from all other comma-bacilli in mode of

I

growth in gelatine, and must therefore be different from all others.

I maintain that notwithstanding the fact that the choleraic comma-bacilli in their mode of growth are not quite so unique as is maintained of them, even when compared with the few species that have already become known—to say nothing about the many species of comma-bacilli that may be discovered if observations are continued and extended—it does not follow that they do not stand in a causal relation to Asiatic cholera. If by clear and definite experiments, imitating as much as possible the methods of infection obtaining in nature, Asiatic cholera can be induced artificially by a pure cultivation of the choleraic comma-bacilli, then it must be considered as proved beyond doubt that they are a *vera causa*, or, in other words, the contagium of cholera.

In a subsequent chapter I shall describe and review all the experiments that have been made in this direction; for the present I shall content myself with the brief statement that such proof is not forthcoming, and that what has been given as proof is highly unsatisfactory.

No true cholera infection, as is understood in pathology and as is proved with some other bacteria, has as yet been produced; all that has been hitherto shown is that the comma-bacilli, like some other notoriously saprophytic bacteria, are capable of producing chemical susbtances acting toxically, or in some animals are capable of producing septicæmia.

One thing, however, may be said with certainty, namely, that as far as our limited knowledge at present goes, in no intestinal disorder in man have comma-bacilli behaving in artificial cultures like those of Asiatic cholera been yet found in the intestinal evacuations. This of course does not mean that no intestinal disorder exists in which the same comma-bacilli are not present to a similar extent as in Asiatic cholera, for, as has been pointed out on a former page, our experience hitherto, and the observations at present available, are extremely limited; but there can, I think, be no doubt, and in this I fully concur with Koch, that in Asiatic cholera comma-bacilli can be with comparative facility detected by the microscope and by cultivation. Hence I agree to the proposition that if in any case of diarrhœa the choleraic comma-bacilli can be shown both by the microscope and by culture-experiments to exist, then the suspicion that it may be a case of Asiatic cholera is quite justified. And it must be clear from this that the discovery by Koch of the choleraic comma-bacilli is, on practical diagnostic grounds, of the utmost importance. For if it should be found that in a locality which is in communication by sea or land with an infected country one or more cases of suspicious diarrhœa had occurred, the demonstration by culture-experiments of the presence in the intestinal discharges of the choleraic comma-bacilli would fully justify us in regarding such cases with grave suspicion, as being probably, though not necessarily, choleraic. At all events sanitary officers, for the sake of the public weal, would be justified in treating these cases as cases of cholera, and in taking measures of isolation and disinfection.

But there is another side not to be lost sight of—it is this. We have seen that not in all cases of marked Asiatic cholera do the comma-bacilli occur in large numbers, since we have mentioned cases, acute and in all respects typical, where the rice-water stools and intestinal contents harbour comparatively few comma-bacilli, scattered amongst crowds of other bacteria. Now in such cases the demonstration by the gelatine-culture test does not invariably yield positive results. The reasons are obvious. For demonstrating by the culture-test the existence of Koch's comma-bacilli there exists at present no better method than that employed by Koch, namely, the gelatine plate-cultivation. Given a mixture of various species of bacteria, and amongst these some species of bacilli and micrococci characterized by rapid growth, and by the capability of rapidly liquefying the gelatine, even if in such a mixture a small number of choleraic comma-bacilli be present, there will be great difficulty in rearing them as colonies in a plate-cultivation. To establish a successful plate-cultivation the number of bacteria introduced into one plate must of course be limited, and the chances are small indeed that more than one or two comma-bacilli are present in a trace of a droplet of a mixture that teems with different species of other bacteria. In such a plate-cultivation, owing to the very slow growth of the choleraic comma-bacilli as compared with many other putrefactive bacteria, and owing to the fact that some of these latter rapidly permeate and liquefy the gelatine, the appearance of the characteristic colonies of the choleraic comma-bacilli will be interfered with. A case like this, then—and it is by no means rare—does not offer very great prospect of success even at the hands of experts in the method, and the microscopic demonstration of comma-bacilli alone can of course be of no diagnostic value, as every one with Koch will readily admit. I have made a series of observations which prove to me that although choleraic comma-bacilli are present in a bacterial mixture, their demonstration by the gelatine culture-test is extremely difficult, or well-nigh impossible. From normal human fæcal matter a mixture was made in 200 to 300 ccm. of distilled water, the mixture being of thickish semi-fluid character. This mixture in every trace of a droplet teemed with bacteria; to it were added ten drops of a pure culture in nutritive gelatine of choleraic comma-bacilli. A series of twelve plate-cultivations was then established after the usual methods of dilution, but in none could any colonies of choleraic comma-bacilli be detected. After a few days in all the plates the gelatine was liquefied and was seething with bacilli and micrococci.

VI.] VALUE OF CHOLERAIC COMMA-BACILLI. 117

In another series, in which to 300 ccm. of fæcal mixture one cubic centimeter of the same choleraic culture was added, out of twelve gelatine plates only in one was there a colony of the comma-bacilli detected. In a third series to 300 ccm. of fæcal mixture 2 ccm. of the pure culture of choleraic comma-bacilli were added; of a series of twelve gelatine plates, two plates showed colonies of choleraic comma-bacilli; one had one colony, one had three out of about thirty colonies of other bacteria. It is evident from this that although crowds of undoubted choleraic comma-bacilli may be present, they are not easy of demonstration if they are present only in comparatively small numbers amongst a great majority of other quickly-growing bacteria.[1] And that this actually obtains in a certain percentage of cases of true Asiatic cholera no one who has had large experience can doubt. I will readily admit that plate-cultivation in a number of cases, particularly those with typical rice-water stools and mucus-flakes, yields positive results if the mucus-flakes are used; but it is well known to all who have had to deal with cholera epidemics that such cases are not by any means common in the early stages of an epidemic, and it is precisely at those stages, when no other symptom besides diarrhœa is present to guide us, and when the bowel discharges are simply fluid fæcal matter, that the detection of the choleraic comma-bacilli would be of the greatest importance. Later on, when the cases have become more numerous, and the symptoms more pronounced and typical of cholera, the

---

[1] Kitisato shows (*Zeitschrift f. Hygiene*, v. 3, p. 487) that the comma-bacilli when kept in a mixture of fæcal matter undergo death: the more rapid, the greater the comparative amount of fæcal matter and the more delayed the examination. This inimical influence of fæcal matter on the comma-bacilli does not affect our argument, since in our cases the examination was proceeded with immediately after the mixture was made. But it helps to explain the disappearance of the comma-bacilli from the stools of cholera patients in the later stages of the disease, pointed out by Koch (*loc. cit.*) and easily confirmed.

diagnostic value by means of the gelatine culture-test is of necessity less important.

The same applies also to cultures made on linen. Koch is easily confirmed when he says that the examination of the mucus-flakes of the bowel discharges, with which the linen of a cholera patient becomes soiled, very often yields a good result, for this represents in some instances what is practically an artificial cultivation. If such mucus-flakes are taken off the linen and kept damp for twenty-four hours they will be found crowded with the comma-bacilli, and their true character can then easily be determined by plate-cultivation. But this result is not always achieved, for if originally but few comma-bacilli and many other bacteria are present, the latter will, particularly in hot weather, have so enormously increased that of the original comma-bacilli very little or nothing is left.

Bujwid's and Dunham's colour test, described on page 111, is also to be mentioned in connection with the diagnosis and detection of the comma-bacilli.

Bujwid[1] shows that the reaction is the more pronounced the purer the culture in choleraic comma-bacilli is; pure cultivations of these bacilli in broth containing pure peptone show after twenty-four hours' growth at $37°$ C. intensive purple reaction with crude HCl.

Gaffky[2] gives an account of an isolated outbreak of a suspicious choleraic disease amongst some of the inhabitants in Gonsenheim and Finthen (near Mayence) during September and October 1886; Dr. A. Pfeiffer had ascertained by microscopic examination and by the culture test the presence, in the stools and in the contents of the intestine of some cases dead of the disease, of comma-bacilli, having all the characters of the choleraic comma-bacilli. Gaffky confirmed this.

[1] *Centralblatt f. Bacteriologie*, IV. p. 494.
[2] *Asbeitur ans d. Kais. Gesundheitsamte*, II. Bad. 1, 2, Heft. p. 39.

CHAPTER VII.

EXPERIMENTAL PRODUCTION OF CHOLERA.

KOCH in his first pamphlet (p. 27) told us that he has made every imaginable effort to produce cholera in animals experimentally. The experiments of feeding white mice with cholera dejecta, first made by Tiersch and then by Burdon Sanderson, were repeated by Koch over and over again on fifty white mice fed with fresh material (dejecta of cholera patients, and the contents of the intestine of cholera corpses) and with choleraic material after it had begun to decompose, but no result whatever followed; the mice remained healthy. "We then made experiments on monkeys, cats, poultry, dogs and various other animals that we were able to get hold of, but we were never able to arrive at anything in animals similar to the cholera process. In precisely the same manner we made experiments with the cultivations of comma-bacilli; these were given as food in all stages of development. When experiments were made by feeding animals with large quantities of comma-bacilli, on killing them and examining the contents of their stomachs and intestines with a view to find comma-bacilli, it was seen that the comma-bacilli had already perished in the stomach, and had usually not reached the intestinal canal. . . . . The comma-bacilli had been destroyed in the stomachs of these animals. . . . The experiment was therefore modified by introducing the substances direct into the intestines of the animals. The belly was opened, and the liquid was injected immediately into the small intestine with a Pravaz's syringe. The animals bore this very well, but it did not make them ill. We also tried to bring the cholera-dejecta as high as possible into the intestines of monkeys by means of a long catheter. This succeeded very well, but the animals did not suffer from it. I must also mention that purgatives were previously administered to the animals in order to put the intestine into a state of irritation, and then the infecting substance was given, without producing any different result. The only experiment in which the comma-bacilli exhibited a pathogenic effect, which therefore gave me hope at first that we should arrive at some result, was that in which pure cultivations were injected directly into the blood-vessels of rabbits or into the abdominal cavity of mice. Rabbits seemed very ill after the injection, but recovered after a few days. Mice, on the contrary died from twenty-four to forty-eight hours after the injection, and comma-bacilli were found in their blood. Of course they must be administered to animals in large quantities; and it is not the same as in other experiments connected with infection, where the smallest quantities of infectious matter are used, and yet an effect is produced. In order to arrive at certainty as to whether animals can be infected with cholera, I made inquiries everywhere in India as to whether similar diseases had ever been remarked amongst animals. In Bengal I was assured such a phenomenon had never occurred. This province is extremely thickly populated, and there are many kinds of animals there which live together with human beings. One would suppose, then, that in this country, where cholera exists in all parts continually, animals must often receive into their digestive canal the infectious matter of cholera, and in just as effective a form as human beings, but no case of an animal having an attack of cholera has ever been observed there. Hence I think that all the animals on which we can make experiments, and all those, too, which come into contact with human beings, are not liable to cholera, and that a real cholera process cannot be artificially produced in them."

I have quoted thus at length from Koch, to show how carefully he had investigated from all points of view the question of the communicability of cholera to animals, and how precise and definite is his conclusion arrived at in this matter. This is the more important since his advocates and imitators leave this entirely out of sight. After Koch had thus declared himself (at the Cholera Conference in Berlin, July 1884), von Pettenkofer—acknowledged to be a high authority on cholera—refused on epidemiological grounds, to be described below, to accept Koch's comma-bacillus as the active cause of cholera; in fact he was so convinced of the contrary that he offered with other medical men, under any conditions to be determined in committee, to swallow any amount of pure cultivation of Koch's comma-bacilli. Fortunately, about that time Nicati and Rietsch had published certain experiments by which they thought they had succeeded in artificially producing cholera in dogs and guinea-pigs. These gentlemen, struck by the fact that in cholera cases with rice-water dejecta the intestinal contents are free from bile, thought the exclusion of bile from the intestine probably formed a *conditio sine quâ non* for the success of the experiment. Any one who has had any experience of a cholera epidemic must be struck with the extravagance of such an idea. Is it not known that many and many a light case of cholera occurs, in which the intestinal contents during the onset of the disease is still mingled with bile, and that as the symptoms increase in vehemence so the bile-secretion, like the other secretions—urine, saliva, gastric and pancreatic juice—ceases? One could understand the reasonableness of such an idea, if the premonitory symptoms of cholera were characterised by suppression of bile-secretion, but this is not the case; most of the secretions of the liver, kidney, saliva, stomach, and pancreas, cease after the disease has well set in. One might just as reasonably suppose that ligaturing the ureter would be a *conditio sine quâ non* for the success of the experiment of artificially producing cholera, on occount of the suppression of secretion of urine in cholera. As a matter of fact Nicati and Rietsch, in their experiments on guinea-pigs, Koch in his experiments on dogs and guinea-pigs, van Ermengem in his experiments on guinea-pigs, did not employ this mode of experimentation.

Well, then, Nicati and Rietsch opened the abdomen of dogs, ligatured the bile-duct, and injected then into the duodenum cultivations of the choleraic comma-bacilla. The animals died; their intestine contained fluid with flakes of detached epithelium; and the comma-bacilli were found to have increased enormously in numbers. This was put down as cholera, due to the increase and action of the comma-bacilli. Was anything more extraordinary ever heard of in experimental pathology? The bile duct is ligatured, the peritoneal cavity and intestine are exposed, inflammation of the bowels sets in, with fluid evacuations and detached epithelium, and this is put down as cholera, and as due to the comma-bacilli introduced. I venture to say that these symptoms can be as readily reproduced without the comma-bacilli. The comma-bacilli are found to have increased in numbers, but surely this should be the natural and inevitable result if they are introduced into an intestine in which disease is set up: such an intestine is their natural breeding-ground; from a diseased intestine they had been originally derived.

Koch repeated (*Deutsch. med. Woch.* 45, 1884) these experiments of Nicati and Rietsch on dogs, without previously ligaturing the bile-buct. The fluid, a fraction of a drop of a cultivation of comma-bacilli greatly diluted, was injected into the duodenum, of course after opening the abdominal cavity. "With few exceptions the animals so experimented upon died after one and a half to three days. The mucous membrane of the small intestine was reddened, its contents watery, colourless, or slightly reddish, and at the same time flakey. In the intestinal contents the comma-bacilli were present in pure cultivation and in enormous numbers. We have then here the same appearances as in the cholera intestine in acute cases." To these experiments I have to apply exactly the same criticism as I applied above to those of Nicati and Rietsch. There is absolutely no guarantee that the peritoneum and bowels of an animal under such an experiment, leaving out the comma-bacilli, would not become inflamed, and in this state the comma-bacilli would readily and copiously multiply. From some experiments made later by Koch, and hereafter to be described, this assumption will appear very feasible.

I recommend to the notice of van Ermengem, and particularly to Mr. Watson Cheyne and Dr. Workman, the following statements on those points made by Koch himself, after a large number of experiments. These gentlemen thought it quite unwarranted and thoughtless on my part to accuse the operation as having anything to do with the result of the experiments. Here is the passage of Koch's that I refer to, as given in his address to the Second Conference on Cholera held in Berlin, May 1885:[1]

[1] *Brit. Med. Journ.* Jan. 9, 1886, p. 62.

"In these experiments it struck me that, the better the operation was performed, and the less extensive the manipulations, the less chance was there of the animals dying of cholera;" and further "in this set of experiments also (without ligature of the bile-duct) the results are so much the less positive the less disturbance and the less the intestine is squeezed or torn in searching for and pulling forward the duodenum. Hence the experiment succeeds only exceptionally, when one limits oneself to opening the abdominal cavity only to a small extent, and making the injection into the coil of intestine first exposed, instead of into the deep-lying duodenum. Of six guinea-pigs which were operated on in this way, only one died of cholera, the rest remained alive. The same experiment was then performed on three rabbits, without any of them dying or even becoming ill."

Well, by these experiments Koch thought himself quite justified in refusing von Pettenkofer's challenge; he said, "Under these circumstances it is more advisable to decline the offer of those persons who recently proposed to swallow pure cultivations of comma-bacilli, and for the present to continue the experiments on guinea-pigs and other animals."

The most exact and carefully-conducted experiments have been made by van Ermengem, and they contrast, as to care and precision and number, as well as in the character of the experiments, with those published by Babes, Doyen, Watson Cheyne, and others. I shall therefore for the present refer only to the experiments of van Ermengem. The most noteworthy result made known by him is this, that guinea-pigs inoculated *per duodenum* with relatively large doses of choleraic comma-bacilli cultivated in serum (one gramme to half a gramme) died in the course of between two and eighteen hours. The symptoms presented by these animals were those of acute chemical poisoning, for what van Ermengem describes on p. 77 of his paper is quite compatible with this. If proof were needed it is furnished by the fact that some of the animals succumbed in a few hours; this could hardly be ascribed to anything else except chemical poisoning, as contrasted with a real infection with an incubation-period such as is known in the case of other infectious diseases. There are now known a good many cases of acute poisoning in the human subject (sausage poisoning, mackerel poisoning, poisoning by over-ripe fruit, tinned salmon, tinned sheep's tongue, &c.) investigated on various occasions within recent years, in which the symptoms of acute gastro-enteric disturbance set in after a few hours. Other symptoms, as the *stadium algidum*, cramps, pain in abdomen, fall of temperature, disturbed respiration, were all present in these cases. Brieger in his pamphlet on the *Ptomaïnes* (Berlin, 1885) has produced these symptoms in animals by the various alkaloids, analysed and isolated by him from various food-stuffs, that had undergone putrefaction. The guinea-pigs thus experimented upon by van Ermengem showed on post-mortem examination of their intestines amongst numerous septic bacilli and micrococci a few of the choleraic comma-bacilli. Two observations made by van Ermengen (*l.c.* pp. 78, 80) deserve special notice. (1.) On keeping the intestinal contents of such a guinea-pig for twenty-four hours in a moist chamber, it became crowded with the choleraic comma-bacilli. (2.) One of several animals which had received fractions of a drop of a serum-culture into the duodenum remained well. This animal, nine days after the inoculation, voided in its stools still living choleraic comma-bacilli; from this it follows that the comma-bacilli, although they must have lived and multiplied in this animal's intestine for nine days, were unable to produce any disease.

Another series of experiments, not less important, were made by van Ermengem as follows: by a Chamberland filter or by heating, he eliminated from his cultures (chiefly in serum) the comma-bacilli themselves, and then injected into the duodenum the remaining fluid only; he used this in various quantities, but in most instances produced results identical with the above, *i.e.* death of the animals with the symptoms just described, which he regarded as indicating cholera. But there was this striking difference, that the symptoms and death were retarded in direct proportion to the quantity injected. These observations are in harmony with those of Nicati, Klebs, and others, who have found that in artificial cultivations of the choleraic comma-bacilli there is present a chemical poison, which in guinea-pigs produces acute poisoning similar to ptomaïne-poisoning. But whether from this it can be concluded that when these comma-bacilli are introduced into the intestine, they, by their multiplication necessarily produce therein the same chemical poison, is open to question; moreover, from van Ermengem's observations, I am inclined to conclude the contrary, namely, that this chemical poison is not necessarily produced by them, and that the comma-bacilli can live and thrive in the intestine without producing any such chemical poisoning.

Another important point is this: the production by the comma-bacilli of a chemical poison setting up, when injected into a dog or guinea-pig, symptoms identical with ptomaïne poisoning, is by no means the exclusive property of the choleraic comma bacilli, since various other septic bacteria have the same power,[1] and this has been proved experimentally by Berdez.[2] This gentleman found in artificial cultivations in broth of Finkler's comma-bacillus, and of the jequirity *Bacillus subtilis*, the same chemical poison. As regards Finkler's

[1] See Brieger, *loc. cit.*   [2] *Brit. Med. Journ.* Nov. 7, 1858.

comma-bacillus I shall return later on to experiments made by Finkler.

Hueppe has by recent experiments shown that the hen's egg is the best and readiest means to obtain this toxic substance, by cultivating within it the choleraic comma-bacilli; already after twenty-four hours a considerable quantity of this substance (cadaverin) becomes available. In this respect there is a decided difference between broth and gelatine cultures and those carried on in egg. Further: not only the choleraic comma-bacilli produce in this condition the toxic substance but also Finkler's and Deneke's comma-bacilli when cultivated in this same medium; and that there exists between these different species a difference only in the quantity of the poison produced.

We see, then, that there can be no question about the presence in certain artificial cultures of the choleraic comma-bacilli (particularly in serum-cultures four days old and in egg-cultures) of a chemical ferment capable of producing acute poisoning in animals, but the symptoms thus produced are comparable with ptomaïne-poisoning; and further about the fact that the production of such a ferment does not appertain exclusively to cultures of choleraic comma-bacilli, but also to other saprophytic organisms.

That this ferment is not present in the mucus-flakes of the cholera intestine, although swarming with the comma-bacilli, that it is not present in sufficient quantities in broth and gelatine cultures or in Agar-agar cultures, I have convinced myself by a large number of experiments. Mucus-flakes from a fresh cholera intestine swarming with the comma-bacilli were injected in considerable quantities (one half to one whole Pravaz syringe) into the small intestine of dogs, monkeys, cats, and rabbits, into the jugular vein, and into the peritoneal cavity, but without any result of poisoning.

Numbers of experiments were made at the Brown Institution by Mr. Dowdeswell and myself, by injecting into the duodenum of dogs and guinea-pigs considerable quantities of cultures, recent and old, of the choleraic comma-bacilli in nutritive gelatine and in Agar-agar, but with no result.

I have also, after the manner of Ferran,[1] injected subcutaneously into guinea-pigs large quantities (3–5 ccm.) of cultures of the comma-bacilli in nutritive gelatine and in broth. Two guinea-pigs received each subcutaneously 4ccm. of a culture of Koch's comma-bacillus in beef-broth kept at 37° C. for three days. The fluid was crowded with the comma-bacilli. No result whatever followed.

Two guinea-pigs were similarly inoculated each with 4 ccm. of nutritive gelatine liquefied by, and crowded with the choleraic comma-bacilli; two other guinea-pigs were similarly inoculated, each with 4 ccm. of nutritive gelatine liquefied by Finkler's comma-bacillus. All four animals were dead in twenty-four hours. The symptoms during life were those of ptomaïne-poisoning. Post-mortem examination showed that the whole of the subcutaneous tissue of the chest and abdomen was dark red and œdematous, the viscera much congested, the spleen small. The heart's blood contained in the first two animals Koch's comma-bacilli, in the two second animals Finkler's comma-bacilli, as was proved by cultivations. Cultivations made of the intestinal contents— fluid grumous mucus—yielded no comma-bacilli. Experiments made with small quantities $\frac{1}{2}$—1 ccm. of the above gelatine-cultures, produced no result. Thus it is seen that gelatine-cultures contain the same chemical ferment, but only in relatively small quantity, since huge doses of such cultures are required to produce the effect, small doses being without effect; and it is further seen that this ferment is not

[1] *Comptes Rendus*, 1885.

exclusively present in the cultures of the choleraic comma-bacilli.

Dr. D. D. Cunningham has published (in "Scientific Memoirs by Medical Officers of the Army of India," Part II. 1886, pp. 1 to 14,) a very interesting series of observations made by injecting subcutaneously in the thigh into guinea-pigs cultivations of comma-bacilli directly derived from a case of cholera. Some of the animals thus experimented upon died in two to three days, and showed symptoms of effusion spreading from the seat of inoculation over the lower half of the abdomen of the same side; peritonitis was present, and a sticky secretion was found on the serous covering of the intestine; comma-bacilli were obtained by cultivation from the subcutaneous effusion, from the peritoneal exudation, from the intestinal contents, and from the cardiac blood. These results are then comparable to septicæmic infection, such as was the case in Koch's experiments on mice, in Ferran's and my own experiments on guinea-pigs.

I have found, what has been also noticed by others, that in a large number of animals broth and gelatine cultures of advanced pedigree when injected in large quantities produce no result, and that therein they contrast markedly with cultures of recent pedigree. I have received directly from Dr. Cunningham of Calcutta fresh cultures of the choleraic comma-bacilli and have been able to compare them in this respect with subcultures that have been kept going for two years and more.

Another important fact capable of throwing some light on this question is this. The cases of what is called ptomaïne-poisoning may be grouped into two distinct classes: the one comprises cases in which an alkaloid, the product of putrefaction and putrefactive organisms, is introduced into the system and produces acute poisoning with gastro-enteritic symptoms (vomiting, purging, inflammation of the intestine, its cavity filled with mucus, cramps, fall of temperature, obstructed breathing, &c.). In these cases the introduction of the chemical poison by the stomach seems ineffective, subcutaneous and intravascular introduction, and (what comes to the same) injection into the small intestine, seems a *conditio sine quâ non*. If this be the case, the gastric juice would have a destructive effect on the chemical poison. In another class of acute chemical poisonings, also set down as ptomaïne-poisoning, and in which the symptoms are very much the same, the introduction of the poison into the stomach is perfectly effective. Such are the cases of poisoning by sausages, meat, fish, jelly, pie, &c. Hardly a year passes in which numerous cases of this kind of poisoning do not occur.[1] While then in the first class of cases we have to deal with ptomaïnes, such as have been investigated by Selmi and others, and particularly by Brieger, being the products of putrefaction, we have in the second class to deal with special fermentative processes, the products of which must be of a different nature from the first, since they are unaffected by the gastric juice.

I have had to do with the investigations of such an outbreak of veal-pie poisoning observed by Dr. Thursfield, of Shrewsbury, and I have shown that there was present a species of *Bacterium termo*, which is incapable of life and multiplication in the normal animal body, but when cultivated at the ordinary temperature (18–20° C.) in nutritive gelatine or in broth, rapidly multiplies and produces a chemical fer-

[1] In both instances the rapidity with which the symptoms set in clearly points to an unorganized or chemical poison, in the general acceptation of the term, as distinct from an organized poison requiring incubation, such as we have to deal with in infectious diseases, where certain bacteria are introduced into the system and by their multiplication and life-action give rise to the symptoms.

ment, which, for want of a better term, I called paraptomaïne; this ferment, when introduced into the stomach of mice, produced acute gastro-enteritis, such as the veal-pie did in the human beings who partook of it.

Considered in this light there seems a striking analogy to exist between the chemical poison produced in certain artificial cultures of the choleraic comma-bacilli and the ptomaïnes produced in putrefactive processes such as have been investigated by Brieger. In both the chemical ferment does not pass unscathed through the gastric juice. Series of experiments have been made by a number of workers by introducing into the stomach cholera stools or the contents of cholera intestine and artificial cultures, but without any result. Numbers of people continually partake of substances (meat, game, &c.) that probably contain, judging from the number of putrefactive organisms present in them, considerable quantities of ptomaïnes, yet no disturbance occurs, while in other cases (mackerel, sausage, &c.) serious mischief is produced; these latter cases cannot be simply due to ptomaïnes produced by putrefaction in the ordinary acceptation of the term. But let the ptomaïnes obtained by putrefactive processes be introduced subcutaneously, by the vascular system or otherwise, and independently of the stomach—and such experiments have been repeatedly made by a large number of workers (see Dr. Brunton's recent work on the *Disorders of Digestion*)—and the result is acute poisoning. And such is evidently also the nature of the chemical poison present in certain cultivations of the choleraic comma-bacilli. That only certain cultivations of the comma-bacilli contain this poison in a concentrated form, while others contain little or none, is, as shown in the experiments by van Ermengem, Hueppe, and others above mentioned, quite in harmony with Brieger's observations, who could obtain some of his ptomaïnes only from certain substances, not from others.

Koch having very likely felt that experiments such as those which he and van Ermengem made after the method of Nicati and Rietsch were not free from objection, inasmuch as they involved severe surgical operations (see p. 124), and inasmuch as they, unlike all other experiments employed in bacteriological research, did not imitate the methods of infection as they occur under natural conditions, devised a method of experiment which, though far removed from the first, was not quite free of the second criticism. Starting from the idea that the comma-bacilli are killed by the gastric juice, and that in order to develop their pathogenic powers they have to get unscathed and living into the small intestine —their natural breeding-ground—it occurred to him that this difficulty might be obviated by first neutralizing or making alkaline the contents of the stomach, and then introducing *per os* the comma-bacilli. He therefore kept guinea-pigs for twenty-four hours without food, and injected then into their stomach *per os* 5 ccm. of a five per cent. watery solution of carbonate of sodium. This does not noticeably injure the stomach, and, as direct observation proved, kept the contents of the stomach in an alkaline condition for three hours. Some minutes (twenty) afterwards he introduced by catheter 10 ccm. of a cultivation of the comma-bacilli in meat-infusion.

The result is noteworthy. Seven guinea-pigs thus experimented upon remained perfectly well; " they were killed after twenty hours, and the contents of their stomach, intestine, and cæcum, were examined by gelatine plate-cultivations. In six of the seven animals, the cholera-bacteria could be demonstrated in the small intestine. The experiment had thus in so far succeeded, that the cholera-bacilli had passed uninjured through the stomach; but they had not set up any disease in the animals." Similar experiments were then made on eight other guinea-pigs. These animals also remained quite healthy. Finally four guinea-pigs were similarly experimented upon (5 ccm. of solution of sodium carbonate, then 10 ccm. of cultivation of the comma-bacilli in meat-infusion); three remained well, the fourth appeared ill next day, looked shaggy and did not eat; on the following day it was very ill; paralytic weakness of the posterior extremities came on, the respiration was weak and slow, the head and extremities were cold, and the animal died in this condition. On post-mortem examination the small intestine was markedly reddened and full of a flakey, watery, colourless fluid. The stomach and cæcum contained a large quantity of fluid. "The examination with the microscope and with gelatine-plates showed that the contents of the small intestine contained a pure cultivation of the choleraic comma-bacilli." "That this one animal only should have died, out of a series of nineteen uniformly experimented upon, suggested some peculiar condition that had obtained in this one animal and as a matter of fact on examination it was ascertained that this animal had aborted immediately before the injection, and on post-mortem examination it was found that the abdominal walls were very flaccid and the uterus still greatly enlarged. This led me to the idea that either the abortion *per se*, or perhaps its unknown cause, had acted on the other abdominal organs, more especially on the small intestine, in such a way as to produce a temporary relaxation with arrest of peristaltic movement; and thus had rendered it possible for the comma-bacilli to remain longer and gain a footing in the intestine." This conclusion appeared to Koch justifiable, inasmuch as by direct experiment he thought he had proved that the contents of the stomach pass too rapidly through the small intestine, and since the comma-bacilli could only unfold their poisonous action, *i.e.* could produce the chemical poison, if they had time to remain there and to multiply. Consequently if they were not delayed on their passage through the small intestine they would not multiply there, and once in the cæcum where the reaction is acid, they would become harmless. To this method of reasoning I must take exception. Koch shows by direct experiment that even twenty hours after injection the comma-bacilli can be recovered from the small intestine in a living state. Now the most important character of all pathogenic bacteria is this, that when introduced into the particular tissues suitable for their propagation they set up their pathogenic power. How is it then, one might reasonably ask, that the comma-bacilli, if even only for a few hours in the small intestine, do not invade in swarms the epithelium and superficial layers of the mucous membrane? Koch does not, and of course cannot, deny that all absorption of the chyle must take place in the small intestine, and since the comma-bacilli are much smaller than the large chyle globules, and are possessed of spontaneous mobility, it follows of necessity that the comma-bacilli can and must readily pass into the epithelium and the superficial layers of the mucous membrane; and since the epithelium and the superficial mucous membrane, according to Koch's own statement and belief, are the suitable nidus for the multiplication and action of the comma-bacilli, all conditions would therefore here exist which are required for their settling down and acting. Add to this that 10 ccm. of a broth culture of comma-bacilli containing millions and millions of comma-bacilli, are subject to absorption by the small intestine for twenty hours (see the above-mentioned observations of Koch), and that such vast crowds of comma-bacilli in a few hours kept at the body-temperature ought to yield a most formidable host of descendants, and grave doubts must arise as to the tenability of Koch's explanation.

But to continue. In order to produce a condition similar to the one in the above single successful experiment on the guinea-pig, Koch injected tincture of opium into the peritoneal cavity after the introduction of the sodium carbonate and the cultivation of the comma-bacilli: this answered well for achieving positive results. Immediately after the administration of the 10 ccm. of the culture of the comma-bacilli, 1 ccm. of German tincture of opium for every 200 grms. of the animal's body-weight were injected into the peritoneal cavity; the animal became narcotized for half an hour, and died after one and a half to three days with the same symptoms as the above guinea-pig; "eighty-five guinea-pigs have been infected in this way with cholera."

Now the following criticisms can, I think, be justly applied to these experiments: (1) According to Koch's own showing it cannot be the narcosis which is essential, even allowing for the present that relaxation of the intestine may have been produced by the intraperitoneal injection of opium-tincture, since alcohol alone was injected by Koch into the peritoneal cavity, and he says that thereby "we were most successful in making the animals susceptible to the cholera infection." (2) Can narcosis of the animal be produced by opium without furthering in the least the process of the experiment? This has been tried over and over again; watery extract of opium is injected into the peritoneal cavity, and narcosis lasting for one hour is produced, but the animals remain well; tincture of opium is subcutaneously injected, the animals fall into narcosis lasting for from forty to eighty minutes, but no result is obtained from the previous introduction of the comma-bacilli; in fact the experiment as designed by Koch was repeated by me on a large number of guinea-pigs, thirty in all, but instead of producing narcosis by injection of tincture of opium into the peritoneum I produced it by intraperitoneal injection of watery extract of opium, or subcutaneous injection of tincture of opium and watery extract of opium—but all in vain. The comma-bacilli used were of recent broth-culture, or of gelatine culture, and were beyond question or doubt the choleraic-comma-bacilli. (3) It is not proved that injection of tincture of opium into the peritoneal cavity produces relaxation of the intestine and arrest of the peristaltic movement; there is no proof given for this by Koch as regards the guinea-pig; on the contrary, there are experiments on record made on the dog, when the result of such injection was quickening of the peristaltic movement.

From all these considerations it appears to me unwarranted to conclude as Koch does that the multiplication of the comma-bacilli in the small intestine, and their fatal action by the chemical products they elaborate, takes place on account of a relaxation and arrest of the peristaltic movement by the opium. Another explanation appears to me much more probably correct. It is this—provided the intestine is first made diseased, either in consequence of slight peritonitis, as was probably the case in the guinea-pig that had aborted, or in the experiments when tincture of opium is injected into the peritoneal cavity, or from other reasons, the comma-bacilli that are present in the intestinal cavity undergo rapid multiplication, and by their chemical products not only increase the disorder of the mucous membrane but eventually poison the animal. And from this I conclude further that a multiplication of the comma-bacilli can and does take place only when the intestine is previously brought into a diseased state. Under this view all Koch's and van Ermengem's results become at once intelligible.

I maintain then that the living choleraic comma-bacilli *per se*, however large their number, when introduced into the small intestine are quite innocuous, but they are rendered capable of great multiplication if the intestine is previously, from some cause or another, diseased. The chemical products of such multiplication act as poisons analogous to the ptomaïnes obtained from other putrefactive bacteria.

That this is the true explanation I find proof in some of Koch's experiments with other bacteria, notably with Finkler's and Deneke's comma-bacilli. With both these organisms on experimenting in the above manner he obtained positive results; not so constantly, it is true, but still he did obtain positive results, not identical, but similar. Of course it is not to be expected that, seeing these are three different species, they would act in the same manner. Finkler[1] published a large series of experiments, in which, with his comma-bacilli and after the method of experimentation employed by Koch, he produced results identical with those gained by Koch with the choleraic comma-bacillus. There can be no doubt, from what has been shown above, that Finkler's comma-bacillus has nothing to do with *cholera nostras*, nor with any other infectious disease, but that it is simply a putrefactive organism. And on the same grounds Koch's comma-bacillus cannot be said, by these experiments, to have been proved to have a causal relation to *cholera Asiatica*, any more than has Finkler's comma-bacillus, or any of the other species of septic bacteria that are capable of producing chemical poisons analogous to ptomaïnes. All that can be said is—provided that conditions

[1] *Ergänzungsheft z. Centralb. f. allg. Gesundh.* i. 5 and 6.

are established by which the choleraic comma-bacilli are enabled to grow and multiply in the intestinal canal, these chemical poisons may be produced. A very instructive and parallel case is found in the so-called typhoid-bacillus. As is now generally held, the experiments published by Fraenkel and Simmonds, in which they maintained to have produced typhoid fever and death in rabbits after injection of large quantities of cultivations of the typhoid bacilli do not prove any real infective action of the typhoid bacillus for the rabbit; it has been conclusively proved that this result is entirely due to certain chemical substances generated by the typhoid bacillus in the cultivations (Sirotinin, Beumer and Peiper). This pathological condition can be produced entirely apart from the bacilli by chemical substances (Typhotoxin) produced by them in cultivations, and as is the case in other similar toxic substances the severity of the abnormal state depends on the quantity injected. Moreover it has been shown by Beumer and Peiper that by injection of a small quantity of the chemical substance a refractory state against an otherwise fatal dose of the same substance can be produced. (Beumer and Peiper, *Zeitschr. f. Hygiene*, ii. I, p. 110.)

Of course it cannot be expected that all septic bacteria will behave, both as regards power of multiplication and particularly as regards chemical products, in the same manner as those mentioned above, for it is well known that some do not do so. Hence experiments with these latter do not yield any result, and cannot, therefore, have any value for testing or controlling purposes. In these respects Koch's comma-bacilli do not attain to the dignity of, or at any rate do not surpass certain notorious saprophytic bacteria, which, occurring in normal putrid substances or the human body, are capable when inoculated in small doses into rodents of producing a true septicæmic infection. Thus certain species of bacterium or micrococcus occurring in the fluid of the human mouth were found by Pasteur and Sternberg to act virulently when subcutaneously injected into rabbits, other examples being the *Bacterium termo* found by Brieger in normal fæcal matter, the bacillus isolated by Bienstock from normal human fæcal matter, and the bacillus occurring in the fæces of milk-fed infants. These, as it were normal saprophytic organisms, are capable on inoculation in very minute doses of producing a true infectious disease, a sort of septicæmia, in various rodents, some in rabbits, others in mice, and others again in guinea-pigs. Gamaleïa asserts [1] that using cultures of the choleraic comma-bacilli after their passage through the guinea-pig, (see above) and infecting pigeons with them in successive series the blood of these becomes gradually the proper breeding ground of the bacilli, and that fatal infection in the pigeons can then be produced by the injection into pigeons of a small dose of such blood. If such be the case this would only prove that a septicæmic virus has hereby been reared. Gamaleïa also states that, as in the case of the experiments of Beumer and Peiper on the typhoid bacillus, by the chemical products of the comma-bacilli immunity can be produced in pigeons against the virulent cultures. Löwenthal [2] found that by carrying on subcultures for some time, the comma-bacilli lose their virulence on mice, but can be acquired again by cultivation in a special medium. Mice infected first with weak cultures are found for a time refractory against virulent culture. I have injected into the pectoral muscle of pigeons several cubic centimetres of a recent broth culture of choleraic comma-bacilli; after 24 hours

---

[1] *Semaine Medicale*, No. 34.
[2] *Ibid.*, No. 35.

the muscle and the blood of the heart were used for establishing a very large number of plate cultivations and also for inoculating with considerable quantities broth in tubes, but in no single instance could any growth whatever be produced, nor did the microscopic examination show any organisms. The comma-bacilli had therefore been killed already in 24 hours in the tissues of the pigeon. The animal showed of course no disturbance of any kind.

From these considerations it follows then : (1) the presence of the choleraic comma-bacilli occurs in dead tissues only, as we have shown above, namely the fluid and mucus-flakes of the diseased and disorganized intestine in cholera, (2) the production by the comma-bacilli as by other notorious saprophytic bacteria (Finkler's and Deneke's vibrios) of sulphuretted hydrogen, (see Hueppe's experiments of cultivation of the comma-bacilli in egg) ; (3) the production by the comma-bacilli of Indol in broth that contains peptone (Salkowski, Bujwid); (4) the absence of comma-bacilli in the living tissue of the intestine or other organs, (5) the septicæmic character of the disease produced by the comma-bacilli in certain rodents,[1] and the toxic character produced by the chemical products of certain cultures alike of choleraic comma-bacilli and other comma-bacilli notoriously saprophytic. From these facts the conclusion seems to me to be justified that the comma-bacilli in these respects do not differ from saprophytic bacteria, and their relation to the causation of Asiatic cholera becomes therefore very doubtful.

[1] For a criticism of the recently published researches of Messrs. Macleod and Milles, see the concluding chapter of this work.

CHAPTER VIII.

THE INFECTIVENESS OF CHOLERA.

IN the foregoing chapters we have pointed out that under certain conditions large numbers of comma-bacilli may pass in the living condition through the stomach into the intestine, without producing serious results in the latter organ unless it be previously diseased. Millions may pass thus through the healthy small intestine, as is shown by the numerous experiments of Koch above quoted, without producing any result whatever. And this fact, I think, disposes of the idea that they can be the cause of Asiatic cholera in the human subject. Can any one doubt, who has reflected on the actual conditions obtaining in an epidemic of cholera, that under natural conditions of infection this cannot be so? Is it not one of the most terrible facts known and constantly observed in cholera epidemics, that in a locality where a cholera epidemic has broken out, young and old, healthy and unhealthy, alike are liable to infection? Is it possible, is it in the least justifiable, to assume that in all these persons the state of the stomach at the time of infection was such that its contents were not of the acidity sufficient to kill the few comma-bacilli—for in natural infection it can only be a question of very minute particles of contagium—

that had got access to it ; or stranger still, that in all these persons the intestine is in a state of disease, thus favouring the settling and multiplication of the comma-bacilli? I think all who have witnessed an epidemic of cholera will agree that such an assumption is out of the question. If such were the conditions under which infection takes place, I am sure cholera would be an extremely rare disease. For even assuming that the comma-bacilli had got entrance into the stomach, say before breakfast, when the stomach is supposed to be perfectly empty, and its reaction supposed to be neutral or alkaline, they would pass unscathed and in a living state into the small intestine ; but in order to multiply in appreciable numbers the intestine itself would have to be in a state of disease, otherwise they could not multiply in it. Such conditions would unquestionably reduce cholera cases to an insignificant number. The fact, repeatedly observed during the epidemics of 1884–1886 in Spain and Italy and France, that in a population of 10,000 or 12,000 inhabitants, in the course of two to three weeks 1,200 to 1,500 people were struck down with cholera, disposes of the above assumption. It is true, and it has been observed over and over again, that when cholera appears in a locality it attacks at first and with obvious predilection those persons whose digestive organs, or whose general health for the matter of that, are weakened or in a state of disease; but this is known to be the case in other infectious maladies also, the unhealthy being as a rule more susceptible to infection than the strong. But when an epidemic has well set in, no such exemption of the strong and vigorous is noticeable ; provided the active contagium is present and has in this state access to their system, it does not matter what their general condition or that of their digestive organs is, they are struck down by the plague. And in this respect there is no line of distinction to be drawn between cholera and other infectious diseases. Other infectious diseases, such as typhus, typhoid fever and relapsing fever, also attack with predilection the poor, ill-nourished, and weak; other infections spread by filth and uncleanliness of person, of air and water, but just as in these, so also in cholera, a healthy state of the intestine does not ensure immunity against infection.

It will probably be argued that once the comma-bacilli have settled in the small intestine and multiplied, the fact will not be disputed that they can produce a chemical poison which by absorption produces the symptoms of cholera, and that this at least can be inferred from the experiments on animals; and it will be further argued that this is exactly what Koch maintains. In other words, owing to the absence of comma-bacilli from the blood and the tissues, and owing to their presence in the alimentary canal only, it must be assumed that they produce there a chemical ferment which being absorbed produces cholera, and such an inference is supported by the positive experiments on animals. While admitting that the comma-bacilli, like some other saprophytic bacteria, are under certain favourable conditions of growth capable of producing a chemical ferment analogous to ptomaïnes, I do not admit that this is applicable to cholera. Certain important considerations previously mentioned, for example, that the comma-bacilli can only multiply in an intestine previously diseased, offer an unsurmountable primary difficulty to this assumption; moreover, there is the fact that the comma-bacilli pass only with great difficulty unscathed through the normal stomach, or through a stomach in which the contents are always acid or perhaps of more than normal acidity as in many cases of dyspepsia; whence it would follow that the number of persons subject to cholera ought to be quite insignificant. But there are not less important difficulties of another kind to be settled. Supposing the comma-bacilli in the intestine of a cholera patient produce this chemical poison, say some kind of ptomaïne, does that mean that before the disease has set in the comma-bacilli have been already so numerous in the intestine as to produce this poison in sufficient quantity to set up the disease? There exists no proof that because in a person already ill with cholera the stools and the contents of the ileum contain numerous comma-bacilli, they were present there in large numbers before the disease set in, and there is no proof that when in a well-pronounced acute case the comma-bacilli are very numerous, their chemical products are being absorbed and thereby produce the symptoms. This would be tantamount to placing the cart before the horse. In some, not by any means in all, cases of acute typical cholera that are brought into the post-mortem room, the mucus-flakes of the lower ileum contain large numbers of the comma-bacilli. Does any one mean to say that this state of things existed before the disease set in, or that such an intestine filled with watery fluid is capable of absorbing anything from its cavity? And how about those cases of acute and typical cholera that are brought into the post-mortem room, in which nothing of the sort obtains, *i.e.* where the intestine, amongst crowds of other bacteria, contains only a few comma-bacilli? If it were true, as was first maintained by Koch, that the invasion of the superficial strata of the mucous-membrane of the ileum and its Peyer's glands is the beginning of the disease, and that here the poison is elaborated by the rapidly-multiplying organisms and absorbed, this aspect of the argument would unquestionably lend considerable support to Koch's view, but this I have clearly shown is not the actual case; an invasion of

THE INFECTIVENESS OF CHOLERA.

the superficial layers of the mucous membrane of the ileum and of the Peyer's glands cannot be demonstrated in a number of typical acute cases. All that can be said, therefore, is this—in some cases the comma-bacilli are found numerously present in the mucus-flakes of the ileum (*i.e.* in dead tissue), in other equally acute and typical cases they are there present only sparingly amongst crowds of other bacteria; further, when they are present in large numbers in the rice-water discharges, *i.e.* when the chemical poison is being elaborated in large quantities, the disease has already well set in, and in this state the intestine is pouring out quantities of fluid and therefore cannot be said to be capable of absorption. From these considerations it seems to me to follow that the presence of numerous comma-bacilli in the contents of the cholera-intestine is a result of the peculiar pathological state of the intestine, just as was the case in the above-mentioned successful experiments on guinea pigs. Supposing, as has been suggested on a former page, that the comma-bacilli are already present in the normal human intestine, but being subject to unfavourable conditions remain very limited in numbers; it may be that in cholera the conditions having become favourable we should find them greatly increased in numbers. True, no one has as yet shown that in the normal human intestine comma-bacilli identical with Koch's comma-bacilli do exist, but then few observers have as yet systematically examined the different species of bacteria that do exist in the normal intestine. Koch and others say that they have by plate-cultivations carefully examined the normal contents of the human intestine and have never found them, but it must be remembered that they failed to find *any* comma-bacilli, and yet we now know that Finkler's comma-bacilli do occur normally. I do not mean hereby to imply that these observers have not carefully examined the contents of the normal intestine by gelatine plate-cultivations; I only wish to point out that a negative result obtained in one set of experiments by one operator does not preclude a positive result being obtained in another set by another operator. Again I have shown on a former page that even when choleraic comma-bacilli are added in comparatively small quantity to a large quantity of bacterial mixture, it is extremely difficult, nay, in many experiments impossible, to recover them by plate-cultivation. I say, therefore, that the presence of a few comma-bacilli even in the normal intestine is not impossible, nor even improbable; and that a good many further observations are required to settle this point. I am quite aware that many are content with the facts already at hand; many say—"Here we have a disease in which it is admitted that a peculiar species of bacteria does occur, that this species has not been demonstrated as yet to exist in either normal or morbid states of the intestine; further, that experiments performed on animals prove that these comma-bacilli introduced into the small intestine can under certain conditions multiply there; death with lesions similar to those of cholera ensuing; this is quite sufficient to show that these comma-bacilli are connected with the causation of the disease Asiatic cholera." I fully admit that these facts cover a good deal of ground and that these arguments are without question of considerable strength. But I do not think we ought to be satisfied with them. I cannot overlook the fact that, notwithstanding the constant presence of the comma-bacilli in the cholera discharges and cholera-intestine, they are only present in dead tissues (fluid and mucus in the cavity of the intestine); I cannot overlook that Asiatic cholera would in this respect make one great exception to all other infectious diseases in which at present a causal relation to definite bacteria has been fully established; I cannot overlook the fact that our knowledge of the different species existing in the normal intestine and in the cholera-intestine is far too incomplete to warrant our assuming anything of so definite a nature; I cannot overlook the extremely great difficulties in harmonizing the conditions under which the positive experiments on animals have been carried out with the actual conditions of infection obtaining in nature; I cannot overlook the great discord that exists between what is known of the comma-bacilli as regards their behaviour in gastric juice, and their extremely limited capability of multiplying in the normal intestine on the one hand, and the fearful susceptibility to cholera infection of healthy and vigorous persons in cholera epidemics on the other.

But these are not the only difficulties. In all bacteriological inquiries referring to infectious diseases the results of such inquiries, if they are to be accepted as well-established, must be in harmony with the well-founded facts discovered by epidemiology; the bacteriological portion of the inquiry is no doubt a very important one, but unless it well harmonizes with well-established other facts not appertaining to bacteriology, it cannot claim our full confidence. Now, I maintain with von Pettenkofer that some important parts of our knowledge concerning comma-bacilli do not harmonize with well-established epidemiological facts. Some of these have been already discussed, others will be discussed now. Some of the facts as to the spread of cholera difficult of explanation are those pointed out by von Pettenkofer in reference to the dependence of epidemics on locality and season. Certain localities appear to enjoy a special immunity against the spread of cholera. Versailles and Lyons, Birmingham and other towns, are notorious examples. Persons have carried infection into these localities from surrounding parts, were taken ill with cholera in Versailles or Lyons, and although many such cases were carried thither yet cholera gained no footing in either town. During the last epidemic in the south of France in 1884–1885 numbers of persons coming from Marseilles and Toulon must have carried infection into Lyons, yet this city remained free of an epidemic. The position of Lyons cannot account for this, for it is not, like Rome or Madrid, situated on high ground, where one might suppose the cholera contagium would not easily lodge, but would be gradually carried downward into a lower situation by the natural drainage from a high level into low ground. Lyons, on the contrary, is, as regards its moisture and its situation in the Rhone valley, as badly off as any notorious cholera locality; its cleanliness, its water-supply, its crowded poorer quarters are not a bit better than those of other big cities not enjoying such immunity. This is only one example. In India there are many such localities known; Dr. Cuningham, the late Sanitary Commissioner with the Government of India, in his most instructive book *On Cholera* (Calcutta, 1885), has mentioned several of them.

Von Pettenkofer has minutely dealt with these facts in his various well-known pamphlets and writings on the dependence of cholera on locality, and it is not necessary to enter further into this question. Now this immunity of a given locality, for instance Lyons, seems to me irreconcilable with the facts known about comma-bacilli. A few cases of cholera imported into this locality would yield innumerable masses of comma-bacilli; the nature and position of such localities is, as compared with other cholera-localities, very favourable for the spread of cholera;

VIII.] THE INFECTIVENESS OF CHOLERA. 149

the habits of the poor, the water-supply, the abundance of filth, all combine to make them a good breeding-ground; the comma-bacilli are known to multiply with enormous rapidity, they have been proved capable of growing and multiplying in almost everything that contains animal and vegetable matter, and yet no cholera epidemic seems to result. The effective method adopted in India of moving the troops out into camp and away from a locality in which cholera has broken out proves the same fact. Often soldiers carry infection into such a camp, are there taken ill with cholera, yet with such exceptions no other cases occur. Millions and millions of comma-bacilli are present in camp, still they do not produce infection.

The same holds good with regard to season. A few cases of cholera occur in Calcutta all the year round; there is hardly a month in the year in which isolated cases do not occur. Yet anything like an epidemic is unknown between June and December. The number of cases begin to rise in December, about Christmas time, steadily go on increasing till March and April, then decrease again. The comma-bacilli are available all the year round; the habits of the natives as regards the use of the water from the tanks for all and every purpose remain the same all the year round. If a case of cholera occurs, say in October, in one of the huts or bustees surrounding a tank, the dejecta invariably find entrance into the tank, for this is the natural sewer of the huts; along the shore of these tanks there is any amount of decaying animal and vegetable matter, and there exists here therefore a good and sufficient nutritive medium for the comma-bacilli. In all tanks the natives of the bustees can be seen at all times and seasons performing their external and internal ablutions, washing their linen and their cooking utensils, and in many cases using the water even for drinking. This latter is not the rule, since many fetch their drinking-water from the hydrants, of which there are many in all parts of the city. But there can be no doubt, as inquiry proves, that the water of the tanks is used also for drinking purposes. And yet isolated cases may occur in one of these bustees during parts of the year without being followed by other cases. Calcutta is not an isolated instance, the same holds good of almost every city in Bengal; in Bombay and Benares, when I happened to be there in September and October 1884, I had ample opportunity of studying these facts, and they have been mentioned in the *Report of the English Cholera Commission*, published by the India Office (pp. 28, 29).

There is not a locality in India, in which, owing to the just-mentioned habits of the natives, cholera, once imported, might not be expected to develop into an epidemic; yet this is often not the case. Cases of cholera have been imported by pilgrims and others coming from an infected locality, and while at a certain time of the year, say from March till June, it led to an epidemic, during other times of the year it did no such thing. The same dependence on seasons as regards Europe is well known, and has been very fully discussed and demonstrated by von Pettenkofer.

The notorious dependence of the spread of cholera on season is, I think, irreconcilable with the facts that are known concerning the comma-bacilli. The comma-bacilli grow and multiply well at all temperatures between $16°$ and $40°$ C. I have had good cultures growing at $16°$ C., and therefore the months of August, September, October, and November in India would be extremely favourable. In the south of Europe March and October, or even February and November, would be quite favourable, yet these are, as a rule, the very months when epidemics of cholera are rare; when they do occur, they occur as a rule between the end of April and October. The epidemic in Egypt in 1883, the very epidemic that preceded those of Toulon and Marseilles in 1884, approached its end by the end of October.

If the comma-bacilli were possessed of the power of forming spores, and if only in this state they were capable of producing infection, one could understand that this formation of spores, as is the case with some other bacilli, might be dependent on certain definite conditions, amongst which might be a certain locality and a certain season. But such is not the case; Koch is very definite about it, and others who have devoted special attention to this point are equally definite; I have in a previous chapter discussed this point in detail, and have explained certain appearances which Hueppe thought sufficient for assuming the existence of spore-formation in comma-bacilli. When spore-formation does occur there is no difficulty in demonstrating it; but in the case of the comma-bacilli such a phase is not demonstrated.

Again, if the comma-bacilli were dependent on some special nutriment obtainable only during certain parts of the year and not in others, or if the habits of the people differed as regards cleanliness, water-supply, &c. in certain parts of the year, a difference in the spread of the disease might be then accounted for: but the comma-bacilli live and thrive wherever and whenever there is nitrogenous material—in fact, they are in this respect conspicuous by their small selective power, they grow in all localities, in all climates, and in all seasons.

Then there is the question of the infective power of the cholera-dejecta. If, as many believe, the fresh cholera-dejecta were possessed of infective power, then it would be quite impossible to understand how it happens that the attendants, nurses, and physicians of cholera-patients, those that handle the cholera-dejecta, and the friends and relatives living in the same room with the sick, remain so often exempt. As von Pettenkofer has pointed out, the fact that when in any locality cholera has assumed the epidemic character and the attendants do become liable to cholera does not prove that they contract it from the cholera-patient; for another explanation, namely, that the cholera-virus has by that time become universally distributed in the locality, is a good explanation. If the fresh cholera-dejecta contained the *materies morbi*, or if, for instance, the comma-bacilli were the cholera-microbes, then one case of cholera should be sufficient to infect all those coming near it. The experiments of Koch prove that the comma-bacilli become inert and dead by perfect drying, *e.g.* when dried in a thin layer on a coverglass or on silk threads; but the particles of cholera-dejecta thrown on the floor, on the bed-clothes, &c., do not become so dry that the comma-bacilli are dead. Indeed this is far from being the case. Every one who is working in a laboratory knows that accidental contamination with micrococci floating about in the air is a constant source of annoyance; expose a layer of sterile nutritive gelatine in a glass dish for a few minutes to free air-contamination, particularly during the summer months, then cover it up; or mix the gelatine first with a little dust taken up from any part of the floor of the room and make with it a plate cultivation and you will find in a few days colonies of microbes—bacilli, and particularly micrococci. Yet micrococci are killed by perfect drying. This proves that those contaminating micrococcus germs in the air and dust were not dry. And the same holds good of the comma-bacilli contained in the particles of the cholera-dejecta, for herein the comma-bacilli do not become perfectly dry, and therefore are not dead. Convalescent and weakly persons in hospitals, side by side with cholera-patients voiding innumerable masses of comma-bacilli on to the bed-clothes and the floor, would have very little chance of escaping infection. Yet this immunity is observed over and over again. In the Medical College Hospital at Calcutta I have noticed this, that cholera patients were placed in the general ward side by side with other patients: the native non-cholera patients, like other natives, eat their meals with their fingers, using no spoons or forks, yet I have not heard of any of these convalescents, or the nurses, or anybody else among the attendants, having become infected with cholera. The same thing has been experienced in London and other places during various epidemics. When in India in any city or village a case of cholera occurs, except in the cholera-season, the disease does not spread, yet amongst the natives the constant and close attendance of the relatives on the sick is notorious; no special precautions against contamination with cholera-dejecta are taken, yet no infection occurs. These are conditions which obtain everywhere, and which have been pointed out and demonstrated over and over again in India during non-cholera seasons. If the fresh cholera-dejecta or if the comma-bacilli were the infective agents, such things could not be. Again, as has been already pointed out, the water of the tanks becomes constantly contaminated with cholera-dejecta and therefore also with abundance of the comma-bacilli, and although the water of these tanks is universally used by the people living around them, yet in non-cholera seasons no spread of cholera occurs. I will here give two such instances that came under my own observation in Calcutta in 1884.

It will be remembered that Koch, while in Calcutta, reported to his Government, the substance of which appeared in the *Englishman* of Calcutta, on the 18th February 1884, that cholera having broken out in one of the bustees surrounding a tank in a suburb of Calcutta, he visited this bustee and found numerous comma-bacilli in its tank. On a second visit, a week later, the epidemic being on the decline, he found much fewer comma-bacilli in the water, and this seemed to him and the *Englishman* to furnish a positive and remarkable proof that these comma-bacilli stood in an intimate relation to the cause of the cholera outbreak. It is known to all who have been in India, and it has been mentioned on a former page, that the natives use the water of every tank, ditch, and pool, however dirty it may appear to a European, for all kinds of purposes,—bathing, washing of mouth, washing of domestic utensils, washing of clothes and linen, not even drinking excepted. This particular tank visited by Koch, is, like most other tanks, surrounded by huts, and is used as a sort of common reservoir into which the evacuations of man and beast, and every kind of domestic filth, find access. That the water of such a tank, around which cholera cases occur, and into which the evacuations of cholera patients find access, and in which the clothes soiled with cholera-dejecta are washed, should contain the same comma-bacilli that are present in the choleraic evacuations is what one would naturally expect, and likewise that the number of these comma-bacilli should be fewer the fewer the cholera cases, *i.e.* the smaller the number of comma-bacilli thrown into the water. But to conclude, as Koch does, that because there are comma-bacilli in the water cholera cases occur amongst the people using the water, and as soon as the number of the comma-bacilli decreases in the water the number of cholera cases becomes less,[1] is manifestly illogical. That Koch should have used an argument of this nature to build up his theory is only intelligible if we remember how little convinced some of the medical public appeared to be of Koch's theory and that it required, as it were, a much stronger argument to support it. This discovery of the comma-bacillus in the water of that tank was considered such an argument, as is clear from the manner in which at the time the daily and some of the medical papers wrote about it.

That the cholera virus, whatever this is, can find entrance into a person by being conveyed there by water is in perfect harmony with the known facts, and that pure drinking water not contaminated with any extraneous material is of the greatest importance finds many a good illustration in the Reports of the Privy Council Office, in the Broad Street Pump cases in London, in Dr. Macnamara's work on Cholera, and in the various Indian Sanitary Reports.

Another curious illustration how even a very experienced observer like Koch sometimes becomes unable correctly to interpret plain facts, is furnished in the same reports sent to the German Government. Koch stated that at Fort William, in Calcutta, cholera abated as soon as a good water-supply to the fort was introduced, and takes this of course as proof that, previous to the introduction of the good water-supply, many cholera cases were due to contaminated water. Now, had he taken the precaution as he might easily have done,

[1] This last statement of Koch's requires a certain amount of correction. The tank of which Koch speaks was visited by him on the 13th February, and again on the 20th February. During the week the comma-bacilli had greatly diminished, but in the records of the police office I find that the epidemic in the bustees surrounding this tank broke out on the 21st of January, and lasted till the 27th of April. It lasted, therefore, fully two months more after this conspicuous diminution of the comma-bacilli.

of looking at the records, he would have found that such a conclusion was not in harmony with the actual facts, for he would have found by studying the records, that cholera cases diminished in a very marked degree some years *before* the introduction of the better water-supply, and that this diminution, but no greater one, was kept up afterwards.

The *Indian Medical Gazette* of November 1884 republished, on page 332, the official statistics as to the course of cholera in Fort William from 1856 to 1876. In 1863 there occurred a sudden decrease of cholera, and this decrease was kept up till 1876. But the new and pure municipal water-supply was not introduced in 1862 or 1863, but in 1872, *i.e.* nine years later than the conspicuous decrease of cholera happened.

I had the opportunity in connection with Dr. D. D. Cunningham to make an examination of the water of some of the tanks in Calcutta, with reference to this very question of the comma-bacilli. The same tank that plays such a conspicuous part in Koch's report above mentioned was visited by us on the 26th November. It is situated in Sahil Bagan, a suburb of Calcutta, and it is surrounded by native huts, in which altogether about 200 families are living. There had occurred one case of cholera in one of these huts about the first week of the month of November. The water of this tank was very dirty, particularly all along the shore, and the people around the tank, as is customary, made use of the water for all and every kind of domestic and other purposes, including drinking.

A sample of this water was taken from near the shore where it appeared particularly impure, about twenty yards from the house in which the cholera case had occurred, and the microscopic examination revealed living comma-bacilli which, as was proved by cultivation, were identical in

VIII.] THE INFECTIVENESS OF CHOLERA.   157

every respect with those found in choleraic dejecta. Notwithstanding their presence in this water, and notwithstanding the extensive use the 200 families were constantly making of it, there has been no outbreak of cholera. Now we have in this instance an experiment performed by nature on a scale large enough to serve as an absolute and exact one. This water had been unquestionably and notoriously contaminated with choleraic evacuations, and therefore also with the comma-bacilli, and was used extensively by many human beings for several weeks; if we say with Koch that the comma-bacilli were the cause and essence of cholera, how is it that not one person amongst so many, up to the middle of December and afterwards, contracted the disease? Clearly because the water did not contain the active cholera virus, and because this latter cannot be identical with the comma-bacilli.

It might be said, and Koch has said so, as a matter of course, in criticising my observations, that perhaps the comma-bacilli present by the end of November were not the same as the cholera-bacilli; but it must be remembered that a case of undoubted cholera having here occurred, owing to the conditions obtaining and owing to the habits of the people, large quantities of comma-bacilli must of necessity have been thrown and carried into this tank; along the shore the water contained abundance of decaying animal and vegetable nitrogenous material to form a very good and suitable nourishing medium for the bacilli, and they must have had ample opportunity to multiply, and consequently there must have been large numbers of them present, sufficient for hundreds of human beings: nevertheless no case of cholera occurred.

An equally striking illustration of the innocuousness of the comma-bacilli is furnished by a tank situated near

Teleepara Lane, in Calcutta. Between the 14th and 16th of November 1885 there occurred nine cases of cholera in three houses of Teleepara Lane, Nos. 3, 4, and 34. No. 34 had three cases, No. 3 had three, and No. 4 had three cases. The people of No. 34 are rich Hindoos, and those of No. 3 and No. 4 are also well-to-do. Two of these three houses have their own hydrant, and from it they have a good supply of very clear water, such as is supplied to all good houses in the town. There is no condition common to all three houses, except that just in front of each of them there appears to be a communication with the street sewer. A narrow passage leads from Teleepara Lane to a bustee surrounding a large tank. As is usual the people (low caste) living in this bustee make extensive use of the water of this tank, but the people of those three houses, being well-to-do and having their own drinking water, never went near this tank. In one of the huts of this bustee lives a milkman, who supplied amongst others, house No. 34 of Teleepara Lane, but not No. 3 or No. 4. The water of the tank, as usual, is very dirty, especially near the shore, and a sample of it examined under the microscope revealed the comma-bacilli. These proved on cultivation to be identical with the choleraic comma-bacilli. How they got there I am unable to say, but it is highly probable that linen from the house No. 34 was washed in the tank; and on inquiry it was ascertained that this was actually the case with the linen of all the houses in the neighbourhood. Amongst the people of this bustee there had not occurred a single case of cholera during the whole year.

It is quite clear from all this that the statement of Koch and his adherents as to the importance of the comma-bacilli in the water in producing cholera is not borne out by these observations.

It must not however be supposed that I mean to question the statements that cholera dejecta have produced infection, or that water contaminated with cholera dejecta has produced cholera. Such cases of infection are well established. Dr. Snow has minutely described one such epidemic, —the noted Broad Street Pump epidemic, and this is only one among many noticed in former and recent epidemics in Europe. As soon as a certain impure water-supply was stopped cholera cases ceased; to such a water-supply—a river or a well—cholera dejecta had probably had access. This question of the importance of drinking-water as a vehicle of contagion may I think be considered settled. But what is not at all settled is the question whether cholera dejecta when fresh have any power to produce infection, or whether some stage or change has to be passed through by them in order to become infective. At any rate sufficient evidence has been brought forward to show that fresh cholera dejecta have not produced cholera even when mixed with water used for various domestic purposes including drinking. And I presume it is this consideration which led Liebermeister in his recent article on cholera[1] to say that the choleraic comma-bacilli must possess the power of forming spores, since only thus can the theory be brought into harmony with the obvious fact that fresh cholera dejecta have often proved harmless. But to this the answer is obvious, viz., that since the comma-bacilli do not possess, as has been shown on former pages, this power of forming spores, and since they have in many cases of direct and natural experiment on a large scale failed to produce infection, it follows that they cannot be the true cholera germ. A similar criticism is applicable to all that is said by Koch and the contagionists with regard to the infective power

[1] *Specielle Pathologie et Therapie*, i. p. 83.

of linen soiled with cholera dejecta. There can be no manner of doubt that cholera infection has been started from linen soiled with cholera dejecta. These instances are notorious and numerous, and are known from former and recent cholera epidemics; they do not require any special discussion. Now Koch says, and others repeat it with greater or less emphasis, that it is easy to show the existence of the comma-bacilli in the mucus-flakes of the dejecta soiling the linen and clothes of a cholera patient even after days and weeks, provided the linen be kept in a more or less damp state, so as not to dry up and kill the comma-bacilli, and the comma-bacilli are the only bacteria that can be at all considered as playing any part in giving infective power to such linen. While fully agreeing with the first part I totally dissent from the second. It is in harmony with the known observations that the comma-bacilli remain in a living state and therefore are capable of multiplication in such linen, but it is absolutely incorrect to say that they are the only bacteria present.

As I have pointed out in former pages, it is extremely rare to find the mucus-flakes of a cholera stool free of bacteria other than the comma-bacilli. As a matter of fact I have not failed to find in them certain small straight bacilli capable of forming spores; von Emmerich and Buchner always found the bacterium which von Emmerich first pointed out and isolated by cultivation. Whether, as is very probable, such linen harbours still other bacteria no one knows. Koch has not thought it necessary to inquire into this, nor have others, who merely repeat what Koch says. It so happens that the comma-bacilli are very conspicuous by their shape—a fact which first attracted Koch's attention to them[1]—and by their mode of cultivation, as he afterwards

---

[1] *Loc. cit.* p. 6.

VIII.] THE INFECTIVENESS OF CHOLERA. 161

ascertained; but surely this does not make them necessarily the more important. Kern described [1] a bacillus under the name of *Dispora caucasica*, which is very peculiar, and which he found in the Caucasus; it is used, as he first thought, as a ferment to produce from cows' milk a peculiar drink, called kephir or hippö. This bacillus is quite peculiar and distinguished from all other bacilli; it is constantly present in such fermented milk, and is constantly present in the material taken from fermented milk and used by the natives to infect fresh milk. Yet, Kern himself afterwards showed that it is not the *Dispora caucasica* at all which is the ferment, but quite a commonplace *Saccharomyces*, which is also always present, and which he did not at first consider to be the ferment, owing to its commonplace characters. Instances of the simultaneous presence of two or more series of organisms in the same materials or tissues not necessarily connected with the cause of the disease are far more numerous than where the disease microbe is the solitary inhabitant.

Koch has stated in his last paper [2] that amongst the hundred and odd medical men attending the special course to study the choleraic comma-bacilli, one gentleman became affected with 'cholerine'; he voided watery stools, and in them the choleraic comma-bacilli were recovered by gelatine plate-cultivation. This Koch considers as proof that the comma-bacilli by careless handling had found access to the intestine of this gentleman, and there multiplied and produced the 'cholerine'. Von Pettenkofer, who was present at this conference, declined to accept this explanation, but persisted in saying that this 'cholerine' might have been

---

[1] *Biologisches Centralblatt*, ii.
[2] *Zweite Conferenz zur Erörterung der Cholerafrage*, Berlin, May 1885.

produced in some other manner, and that the presence of the comma-bacilli in the stools was a result and not a cause of the state of the alimentary canal. After reading the details of the case, I am quite of the opinion of von Pettenkofer. The gentleman in question had been five days in Berlin when he was attacked by slight digestive disturbance, accompanied with diarrhœa. Surely it is not a very remarkable occurrence, that a gentleman coming from some healthy country place to a big city like Berlin should, under altogether new conditions of water, diet, and mode of life, be attacked by digestive trouble and diarrhœa. To put that at once down as due to his having swallowed a few comma-bacilli is going a little too far. A person thus altered may, apart from other considerations, accidentally have swallowed in Koch's laboratory the choleraic comma-bacilli, which, arriving in a diseased intestine, found the necessary conditions of multiplying; but the disease itself probably was antecedent. Amongst the hundreds, nay, thousands of persons who, in the various laboratories in Europe and elsewhere, ever since the first epidemic outbreak of cholera in France in 1884, have handled the choleraic comma-bacilli in large quantities for the sake of study and experiment, has there been any other case? No; and the above instance, quoted by Koch, is therefore single and isolated; the explanation of 'cholerine' in this gentleman as given by Koch remains therefore unsupported; and it would be going far beyond the necessities of the case if we were to accept the belief of Koch as to the method of infection in this case. That a pathological state of the intestine has a good deal to do with the multiplication of comma-bacilli I have proved by direct experiment. In a monkey, which had received the previous day a dose of castor-oil, and had diarrhœa therefrom, the abdomen was opened under the spray, a loop of the

VIII.] THE INFECTIVENESS OF CHOLERA. 163

lower ileum, just above the ileocæcal valve, and about 4—6 inches long, was ligatured above and below, care being taken not to include in the ligatures the large vessels. With a Pravaz syringe a droplet of mucus was withdrawn from the interior of the loop, and on examination no comma-bacilli could with certainty be discovered. With another syringe about 2 ccm. of a saturated solution of magnesium sulphate was then injected, the loop replaced, and the wound stitched up and dressed antiseptically, the whole operation being done under the spray. Immediately afterwards the animal received subcutaneously one gramme of chloral hydrate dissolved in one to two ccm. of distilled water. This whole experiment was done after the method first employed by Moreau, and repeated by the Committee of the British Medical Association on Cholera.[1] Our animal was killed after 48 hours; on post-mortem examination the ligatured loop was found much injected, its cavity filled with and distended by mucus, containing streaks of blood and numerous flakes. On microscopic examination these flakes contained, besides amorphous mucus and detached epithelial cells, longer or shorter straight thickish bacilli, single or in dumb-bells; these were more or less pointed at the extremities, and many of them included an oval bright spore. There were present numerous comma-bacilli, some single, others in dumb-bells, either S-shaped or with the curve in the same direction, *i.e.* like the outline of a bird on the wing; many were spirals of three or four turns. In some places these comma-bacilli were so numerous and crowded together that the material looked almost like a pure cultivation of them (see Fig. 38). On microscopic comparison it was found that they were of the same character as the choleraic comma-bacilli, except

[1] See the reprint of the Report of the Committee in the *Practitioner*, 1884.

perhaps that they looked a trifle smaller than those in the choleraic mucus-flakes. Cultivations were made with them in six gelatine plates, and in one of these after three days there were no doubt a few colonies which corresponded with those of the choleraic comma-bacilli; this was proved to be the case after two more days. Cultivations in gelatine tubes and Agar-agar tubes yielded growths indistinguishable from the cholera comma-bacilli. In the other plate-cultivations the gelatine was found on the second day liquefied and crowded with the above-mentioned straight bacilli.

FIG. 38.—SPECIMEN OF MUCUS-FLAKES FROM A MONKEY.
1. Spiral forms of comma-bacilli.
2. Couples of comma-bacilli.
Magnifying-power 600.

In another monkey, in which the above experiment had been repeated, the animal died during the second day; the loop was found much distended, and filled with watery fluid, bright red, and containing mucus-flakes. On microscopic examination a few comma-bacilli of the same appearance as those in the first animal could be discovered amongst crowds of straight bacilli. Plate-cultivations did not succeed, the straight bacilli multiplied too rapidly, and in the course of twenty-four to forty-eight hours the gelatine became quite liquefied. Four other monkeys experimented on in the same manner yielded no results; neither microscopically nor by cultivation could comma-bacilli be detected. I conclude from the above successful experiment that owing to the pathological process set up in the intestine, the comma-bacilli, present already before the operation, but in far too few examples to be recognised in the microscopic specimens, had so rapidly multiplied that their demonstration was then comparatively easy.[1]

[1] Messrs. Macleod and Milles, in their paper on Asiatic Cholera, say on p. 168, in reference to the above positive experiment in the monkey: "Klein's experiment itself, if it proves anything, proves that he was dealing with an example of spontaneous generation! Farther, it is not quite clear whether Klein claims that he was dealing with Koch's organism, or only with one identical as to characters given."

From this it is quite clear to my mind that these observers missed altogether the drift of my argument, or did not read the above concluding passage.

Other criticisms made by these gentlemen will be dealt with at the conclusion of the next chapter.

CHAPTER IX.

OTHER BACTERIA IN CHOLERA.

Koch does not describe other bacteria, for he does not think them of any importance; the only ones which he considers important are the comma-bacilli, and on these he first fixed his attention on account of their shape and because in acute pure cases they were in the majority.

In the small intestine, and particularly about the ileocæcal valve, one finds in acute cases of cholera, dissected immediately or very soon after death, freely-floating glassy-looking clumps of mucus, which slightly differ from the ordinary epithelial flakes detached from the surface of the mucus membrane or floating in the clear fluid. They resemble clumps more than flakes, and are more transparent; when examined under the microscope they prove to consist chiefly of mucous or lymph corpuscles, and of a few epithelial cells embedded in a hyaline mucous matter. But the same lymph-corpuscles may occur also, only not so numerously, in the ordinary flakes. These lymph-corpuscles are always numerously present in those peculiar clumps, provided the examination is made *very soon after death*. After an hour and a half or two hours one misses them, since they easily become macerated and disintegrated in the intestinal fluid.

CH. IX.] OTHER BACTERIA IN CHOLERA. 167

They can be found also amongst the flakes of the rice-water stools, provided these are quite fresh, but then they are obtained only in a fragmentary state. But the sooner the post-mortem examination is made the more numerously they are found in those glassy clumps. Lewis and Cunningham in their reports on cholera have noticed them, and they correctly state that in order to see them the material must be fresh, *i.e.* examined very soon after death. One misses any mention of them in Koch's paper, whether it be that his attention was chiefly or wholly directed to the comma-bacilli, or, what seems more probable, his dissections were not made sufficiently soon after death. That this is the more likely explanation appears from the fact that when stained with aniline dyes many of these corpuscles contain some interesting things, as will appear presently; and had those corpuscles been present in Koch's specimens he could not have failed to notice their contents. Examining these mucous corpuscles in preparations dried (after the Weigert-Koch method in thin layers) and stained with gentian-violet, or Spiller's purple, or methyl-blue, they present themselves as spherical, oval, or irregular corpuscles of about the diameter of ordinary white blood corpuscles, or larger, if swollen up. Each contains two or three deeply-tinted oval, spherical, or angular nuclei. Their protoplasm is more or less hyaline, and they vary in size, inasmuch as many of them show signs of being swollen up or are even in the act of disintegration, as is indicated by their faint or broken outline respectively. The best preserved spherical corpuscles are completely filled with very *minute straight bacilli*. Those that are slightly swollen show the bacilli more isolated, but still in many places in groups, and in those that are much swollen up and at the point of disintegration the bacilli are seen very loosely and irregularly scattered through the protoplasm or on the point of leaving the corpuscle altogether. The accompanying figures (39 and 40) illustrate all these points. In the surrounding fluid one always meets with the same minute bacilli scattered about. The appearances presented by these mucous copuscles filled with the bacilli and by those that have swollen up and in which the bacilli are loosely scattered, are extremely striking, since the bacilli are stained deeply, whereas the cell-substance appears homogeneous. These lymph-corpuscles are always to be met with in the glassy clumps and under the conditions mentioned above; but not in all instances does

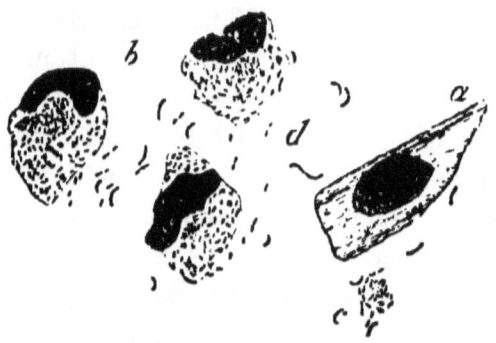

Fig. 39.—From a Preparation of Fresh Mucus-Flakes from the Ileum of a Typical rapidly fatal Case of Cholera.
(a) An epithelial cell.
(b) Lymph corpuscles filled with the minute straight bacilli.
(c) A mass of small bacilli and a few commas.
(d) Comma-bacilli.

Magnifying power about 700.

one find that they contain the same abundance of the small bacilli, for in some cases these latter were missed in most of the well-preserved corpuscles, and found only in those that had slightly swollen up or were on the point of disintegration. But in all instances the same small bacilli are found scattered amongst the detached epithelial and lymph cells. There has not been a single case examined in which they were not found in the mucus-flakes; in cases in which the comma-bacilli were very scarce, the small bacilli were not scarcer. In most cases they were met in larger or smaller groups and as isolated examples.

[As one amongst several cases interesting as regards the occurrence both of comma-bacilli and the small straight bacilli is the following :—E., aged 25, had been purging and vomiting since twelve o'clock in the night of 15th November; was admitted into the Medical College Hospital,

Fig. 40.—From a Preparation of fresh Mucus-Flakes from the Ileum of another Typical rapidly fatal Case of Cholera.
Lymph corpuscles containing the minute straight bacilli.
Magnifying-power about 700.

Calcutta, on the 16th November, at 10 a.m., with symptoms typical of the acute stage of cholera. Died at 1.45 p.m., *i.e.* a little over twenty-five hours after the first attack. Post-mortem examination at 2.20 p m. Ileum contains clear watery fluid, with glassy mucus-flakes, and numerous epithelial flakes. In the mucous membrane a few minute hæmorrhagic spots, not bigger than the point of a pin; Peyer's glands not visible.

In the mucus-flakes were large numbers of lymph-corpuscles, some perfect and small, others swollen up; many of them contain the small straight bacilli in great numbers; besides these there were numerous coherent masses entirely composed of the small bacilli, but comma-bacilli were also everywhere to be found, though the small bacilli were in the majority. Cultivations made on linen from these mucus-flakes yielded after twenty-four hours large crops both of comma-bacilli and of the small straight bacilli.]

These bacilli are of extremely small size, about half to two-thirds the thickness of the typical comma-bacilli, and about one-third their length. They are straight and appear pointed at each end; generally they are single, but occasionally they form a chain of two elements. In the well-preserved mucus corpuscles they lie closely packed together, apparently all single; in the large swollen corpuscles there are some in couples; and amongst those occurring free around and between the lymph-corpuscles and epithelial cells there are a good many in couples and in small groups. It is not at all a rare occurrence to meet with mucus-flakes from rice-water stools in which the corpuscles were found almost completely disintegrated; there were nevertheless found many groups of the small bacilli, from six to twenty and more in each group.

Two questions present themselves in connection with these lymph-corpuscles; (1) where do they come from? and (2) where do they get the bacilli from? There can be no difficulty in answering the first. It is well known that in all those places where the highly-vascular lymphatic tissue reaches the free epithelium of a mucous membrane, *e.g.* the tonsils of the palate and pharynx, the lymph-follicles of the pyloric end of the stomach and the duodenal part of the intestine, the solitary and agminated lymph-follicles of the ileum, and those of the Peyer's glands of the lower part of the ileum and ileocæcal valve, lymph-corpuscles pass (migrate) easily through the surface epithelium, and are discharged on to the free surface. This is the case in the normal condition to a certain extent, and to a greater extent in the pathological state. The mucous corpuscles found in the fluid of the mouth are those that have passed out from the superficial lymphatic tissue of the tonsils. In the Peyer's glands of the ileum one constantly meets these same corpuscles on their way through the epithelium of the surface.

The second question is more difficult to answer. From the fact that the bacilli are found inside and outside the mucous corpuscles, it might be said that the mucous corpuscles being endowed with amœboid movement while, and immediately after, passing out of the mucosa, are probably capable of swallowing the bacilli just as lymph-corpuscles are capable of swallowing other granular matter; but against this might be urged, that the mucous corpuscles, having passed out of the mucosa, probably do not long retain their amœboid power; proof being afforded by the rapidity with which they swell up and become disintegrated in the watery contents of the intestine. The fact that the better the corpuscles are preserved the more numerous the bacilli, might be an argument either way; and, besides, several cases have been examined with this view, and only in one were the bacilli found plentiful within the well-preserved corpuscles: they were absent, or almost absent, in the well-preserved corpuscles of other cases, but were present in small numbers in those that had already swollen up or commenced to disintegrate. There is one other point which must be mentioned in connection with this,—it is the fact that, although these bacilli are not endowed with locomotion, it is not impossible that they settle on these corpuscles, and penetrate by active growth into them, finding in their protoplasm a good soil.

A very careful examination of fine microscopic sections of different parts of the intestine, well-preserved and well-stained in the different aniline dyes, was made in order to trace, if possible, these small bacilli, isolated or enclosed in cells, from the lymphatic tissues of the mucous membrane outwards, but all in vain. No trace of them could be found in the lymph-corpuscles or any other part of the mucous membrane either in the stomach, intestine, mesenteric glands, blood, or any other tissue.

On the whole, then, although these bacilli looked very promising at first as regards their connection with the disease, they had nevertheless to be abandoned, and had to be regarded like the comma-bacilli, as something extraneous, present only in tissues practically dead in the cavity of the alimentary canal. But if any one wishes to urge that these small bacilli are probably connected with the disease, there would exist for such a view at least as much, if not more, justification than for Koch's comma-bacillus, since these small bacilli are found in some elements derived from the tissue of the intestine (the comma-bacilli are not), and are always present in the mucus-flakes and in the intestinal contents, at any rate in acute cases, and if post-mortem examination be made soon enough, as often and as numerously as the comma-bacilli. In the watery vomit, when copious, of acute cases, these small bacilli are generally present, chiefly as isolated individuals or in small groups.

And in the same way, one might further urge that they are quite capable of forming some kind of chemical ferment, which, when absorbed, produces the disease. All this could be said, with the same justification, of these small bacilli as Koch has said it of the comma-bacilli, and such a theory would rest on a basis not a bit weaker than the one on which Koch's theory of the comma-bacillus rests.

These small bacilli have been cultivated in the same way as the comma-bacilli, on linen kept moist by filter paper under a bell-glass, on mixtures of Agar-agar, meat extract and peptone, alkaline and neutral, and their characters have thus been studied. They grow well at ordinary temperature (75° to 82° F.), so that after twenty-four to forty-eight hours considerable masses become available; of course they grow much more rapidly at higher temperatures (90° to 102° F.), and they grow like the comma-bacilli and other bacilli much better and more copiously in alkaline than in neutral media.

The appearances presented after inoculation with them of Agar-agar material in test-tubes are very much like that presented by the comma-bacilli: from the point of inoculation the growth spreads in the form of a flattened or filmy rounded whitish mass, its outlines uneven or knobby. Preparations made of culture on linen and on Agar-agar mixture (solid), show the bacilli singly or very often in chains of two or dumb-bells; the single bacilli are of the same small size as those mentioned above, but many of them grow to somewhat greater length in the cultivation than in the fresh material. After twenty-four to forty-eight hours' growth (at 90°—102° F.) some of them begin to show the formation of spores in the shape of a bright glistening spherical granule, the substance of the bacillus gradually becoming pale, not staining, and ultimately altogether fading away, so that only the spore is left. After several days' growth many of the bacilli, which have not formed spores, become pale, stain very faintly, and gradually fade altogether away. This change, indicating the degeneration and death of the bacilli, differs in no way from what was observed of the comma-bacilli, and described on a former page. Growing in nutrient gelatine they do not liquefy the material, and the channel of inoculation after several days' growth is occupied by streaks of granules and droplets of a whitish appearance.

Experiments were made with cultures of these small bacilli on monkeys, rabbits, dogs, and cats, by feeding, by intravascular and subcutaneous injection, and by introducing them directly into the small intestine. But no result was produced and no result was to be expected, since the experiments with the mucus-flakes taken directly from the ileum of acute cholera cases mentioned on a former page proved without result.

Von Emmerich stated that in cases of cholera examined by him in the epidemic in Naples in 1884, he found a bacillus which is constantly present in the dejecta, in the tissue of the intestine, liver, spleen, lymphatic glands, and blood. This bacillus was found to be virulent when inoculated from cultivations into guinea-pigs, producing death in a day or two with choleraic symptoms. In his later researches carried on with Buchner in Palermo in 1885, he corrected some of his original statements in so far that the presence of this "cholera-bacillus" in the blood and tissues was not confirmed. But its constant presence in the stools and intestinal contents of acute cholera cases and in the mucus of the bronchial tubes, as also its virulently poisonous action on guinea-pigs, was maintained by these observers. I first thought that von Emmerich's bacillus was the same as the minute straight bacillus described by me, but from further more detailed description and from information given me by my friend Dr. Shakespeare of Philadelphia, who had seen and possessed specimens of von Emmerich's bacillus, it is clear that this latter is much larger and thicker than the minute straight bacillus mentioned by me. At the same time von Emmerich's bacillus appears more like a species of *Bacterium termo*, and as such is also regarded by Koch and others. Although I do not of course doubt that this bacterium does occur in the contents of the cholera intestine and the choleraic dejecta, and although there can be no doubt that it possesses those pathogenic characters on guinea-pigs which are stated by von Emmerich and Buchner, yet I cannot for one moment accept it as proved that it has a causal relation to *cholera Asiatica*. Koch and Brieger maintain that the same bacterium occurs in the normal intestinal contents, and the latter observer and Weisser[1] have proved that a bacterium identical with von Emmerich's in morphological and cultural characters occurs in normal human fæces and other localities, and is possessed of the identical pathogenic properties on guinea-pigs, these animals after inoculation dying from a form of septicæmia. All the difficulties that the comma-bacillus of Koch offers in trying to explain the known facts of cholera are likewise attached to this bacterium of von Emmerich's, and I quite agree with those who say that of the two Koch's comma-bacillus has undoubtedly a stronger claim to be considered as the cholera microbe than von Emmerich's. Of course if it had been confirmed that von Emmerich's bacterium is present in the blood and tissues of acute cholera cases there would have been strong *prima facie* evidence for its being causally connected with cholera, but this presence in the blood and tissues not having been proved on further examination its claim to be considered as the cholera microbe rests on a very slender basis.

A commission consisting of Professor C. Roy, Dr. Graham Brown, and Dr. Sherrington of Cambridge, was sent out to Spain in 1885, to decide between the contradictory statements as to the facts concerning the comma-bacilli of Koch. These gentlemen have come to the con-

---

[1] *Zeitschr. f. Hygiene*, i. 2.

clusion that the comma-bacilli are not constant in cholera. They in their Report (printed in the *Proceedings of the Royal Society*, No. 247, p. 173) state that they are unable to accept the comma-bacillus of Koch as causally connected with *cholera Asiatica*. They look upon the comma-bacilli as probably connected with the premonitory diarrhœa; but these gentlemen furnish no proof for this assumption. Messrs. Roy, Brown, and Sherrington describe and figure in sections through the mucous membrane of the cholera intestine preserved for some months hyphæ or mycelial threads which they were told by Mr. Gardiner were the hyphæ of *Chitridiaceæ*, and they are not disinclined to look upon these as causally connected with *cholera Asiatica*. I have good reasons for saying (see *Nature* for December 23, 1886, and the *British Med. Journal* of December 25, 1886) that what these gentlemen figured and described (in *Proceedings of the Royal Society*, No. 247) are the hyphæ of common mould which must have grown into the tissue during the process of preserving the material.

It is fair to state that Mr. Gardiner has subsequently (*Nature*, January 20, 1887) altered his view, inasmuch as he considered the organism shown to him in Professor Roy's specimens, *i.e.* moniliform threads with terminal nodular swellings, to resemble an involution form of a bacterium. Still later (*Nature*, February 3, 1887) he implied that to harmonise what he saw in Professor Roy's specimen with what has been figured by Roy, Brown, and Sherrington in their Report (*Proc. Roy. Society*, 247, p. 173), *i.e.* distinctly branched mycelial threads, both might belong to a form similar to *Cladothrix dichotoma*. I have not the least doubt from actual observation that the branched mycelial threads figured in the Report of Messrs. Roy, Brown, and Sherrington are threads of common mould.

Such appearances cannot be found in sections through the cholera intestine preserved under the necessary precautions in alcohol, as for instance if small bits of the intestine taken out soon after death are placed at once in a large quantity of strong alcohol. As a matter of fact Messrs. Roy, Sherrington, and Brown missed these forms in cover-glass specimens made of the contents of the cholera intestine. Messrs. Sherrington and Rouse, who studied cholera in Italy in 1886, failed to find any of these hyphæ in the cholera intestine. Dr. Shakespeare of Philadelphia studied in 1886 cholera in Spain and India, and in his Report to his Government, arrived at the conclusion that while Koch's comma-bacilli are always present in the early stages, and therefore are of great diagnostic value, their causal relation to cholera has not been satisfactorily established.

In a paper, "Abstract of the Results of an Inquiry into the Causation of Asiatic Cholera" (reprinted from the *Proceedings of the Royal Society of Edinburgh*, and published in the *Reports* from the Laboratory of the Royal College of Physicians, Edinburgh, vol. i. p. 161), Messrs. Neil Macleod and Walter J. Milles state that they have repeated Koch's experiments, and have arrived at the same results. In criticizing my statements they make certain strictures on me which are quite unwarranted.

On p. 173 they say: "Klein's experiments, in which he gave the opium in other ways than by the peritoneal cavity, and then injected the cholera-bacillus with negative results, are inconclusive, as he never made a control experiment with his material to see whether he was able to produce Koch's results under Koch's conditions." This is in so far an unwarranted statement, as I have made control experiments under Koch's conditions with the results described by Koch. After the eighty-five successful experiments recorded by Koch, I do not see the necessity for me or any one else to emphasize the correctness of Koch's observations. What I wished to point out was, that in order to achieve those positive results, viz. multiplication of the comma-bacilli in the intestine and consequent death of the animals, it is necessary to induce a diseased state of the intestine, and this is done by injecting the *tincture* of opium into the peritoneal cavity. For I showed that if the narcosis be produced otherwise—*e.g.* by subcutaneous injection of opium tincture or watery opium extract, or by intraperitoneal injection of watery extract of opium—no result follows the introduction of cultures of comma-bacilli into the intestine after soda injection. So that, provided the intestine be not injured, which assuredly it is by the injection of tincture of opium or alcohol into the peritoneal cavity, the comma-bacilli are unable to multiply in the intestine, and therefore are unable to produce a fatal result.

But while I have made the necessary control experiments which Messrs. Macleod and Milles charge me with having omitted, I do not find anywhere in Messrs. Macleod and Milles's paper a reference to control experiments of their own. Amongst the numerous experiments made by these gentlemen, some partly in repetition of Koch's, partly of Van Ermengem's experiments, I do not find any experiments to show that my contention is wrong. Surely if any one maintains, as they do on p. 177, 3, that "the means used to introduce the comma-bacillus into, and those used to lessen the peristalsis of the small intestine of the guinea-pig, cannot be regarded as causing appearances like those of Asiatic cholera, or as causing the death of the animal,"—we should require proof by control experiment, *i.e.* we should require proof that by inducing narcosis otherwise than by intraperitoneal injection of opium tincture we nevertheless obtain the same positive results. But such control experiments do not seem to have been made by them.

Nor do I find anywhere in Messrs. Macleod and Milles's paper a reference to the important experiments made on a large scale by Finkler with Finkler's comma-bacillus, and described in a former chapter (p. 137). As has been stated, Finkler experimenting on guinea-pigs with cultures of his comma-bacillus after Koch's method produced results identical with those produced by Koch's comma-bacillus. I think this proves conclusively that the action of the choleraic comma-bacillus thus experimented with in the guinea-pig cannot possibly be said to be identical with *cholera-asiatica*; for if so, then both Finkler's and Koch's comma-bacillus would have to be regarded as the cause of cholera, which would be an absurdity.

In criticizing an argument of mine as to the exemption from cholera of attendants and those who constantly are brought in contact with cholera dejecta, Messrs. Macleod and Milles compare on p. 177 the mode of spread of the cholera contagium with that of syphilis; this, I think, is scarcely necessary for me to seriously consider.

www.ingramcontent.com/pod-product-compliance
Lightning Source LLC
Chambersburg PA
CBHW032134160426
43197CB00008B/634